OUR SOLAR SYSTEM

OUR SOLAR SYSTEM

A W JOSHI
Former Professor of Physics
University of Pune
Former Hon. Visiting Scientist
National Centre for Radio Astrophysics
Pune, Maharashtra

N C RANA
Formerly in Astrophysics Group
Tata Institute of Fundamental Research, Mumbai
(Later at Inter-University Centre for Astronomy and Astrophysics
Pune University Campus, Pune)
Maharashtra

NEW AGE REFERENCE
An Imprint of
NEW AGE INTERNATIONAL (P) LIMITED, PUBLISHERS
LONDON • NEW DELHI • NAIROBI
Bangalore • Chennai • Cochin • Guwahati • Hyderabad • Kolkata • Lucknow • Mumbai
Visit us at www.newagepublishers.com

Published by New Age International (P) Ltd., Publishers
First Edition: 1992
Reprint: 2017

GLOBAL OFFICES

- **New Delhi** **NEW AGE INTERNATIONAL (P) LIMITED, PUBLISHERS**
 7/30 A, Daryaganj, New Delhi-110002, (INDIA)
 Tel.: (011) 23253771, 23253472, **Telefax:** 23267437, 43551305
 E-mail: contactus@newagepublishers.com • Visit us at www.newagepublishers.com
- **London** **NEW AGE INTERNATIONAL (UK) LTD.**
 27 Old Gloucester Street, London, WC1N 3AX, UK
 E-mail: info@newacademicscience.co.uk • Visit us at www.newacademicscience.co.uk
- **Nairobi** **NEW AGE GOLDEN (EAST AFRICA) LTD.**
 Ground Floor, Westlands Arcade, Chiromo Road (Next to Naivas Supermarket)
 Westlands, Nairobi, KENYA, **Tel.:** 00-254-713848772, 00-254-725700286
 E-mail: kenya@newagepublishers.com

BRANCHES

- **Bangalore** 37/10, 8th Cross (Near Hanuman Temple), Azad Nagar, Chamarajpet, Bangalore- 560 018
 Tel.: (080) 26756823, **Telefax:** 26756820, **E-mail: bangalore@newagepublishers.com**
- **Chennai** 26, Damodaran Street, T. Nagar, Chennai-600 017, **Tel.:** (044) 24353401
 Telefax: 24351463, **E-mail: chennai@newagepublishers.com**
- **Cochin** CC-39/1016, Carrier Station Road, Ernakulam South, Cochin-682 016
 Tel.: (0484) 2377303, **Telefax:** 4051304, **E-mail: cochin@newagepublishers.com**
- **Guwahati** Hemsen Complex, Mohd. Shah Road, Paltan Bazar, Near Starline Hotel
 Guwahati-781 008, **Tel.:** (0361) 2513881, **Telefax:** 2543669
 E-mail: guwahati@newagepublishers.com
- **Hyderabad** 105, 1st Floor, Madhiray Kaveri Tower, 3-2-19, Azam Jahi Road, Near Kumar Theater
 Nimboliadda Kachiguda, Hyderabad-500 027, **Tel.:** (040) 24652456, **Telefax:** 24652457
 E-mail: hyderabad@newagepublishers.com
- **Kolkata** RDB Chambers (Formerly Lotus Cinema) 106A, 1st Floor, S N Banerjee Road
 Kolkata-700 014, **Tel.:** (033) 22273773, **Telefax:** 22275247
 E-mail: kolkata@newagepublishers.com
- **Lucknow** 16-A, Jopling Road, Lucknow-226 001, **Tel.:** (0522) 2209578, 4045297, **Telefax:** 2204098
 E-mail: lucknow@newagepublishers.com
- **Mumbai** 142C, Victor House, Ground Floor, N.M. Joshi Marg, Lower Parel, Mumbai-400 013
 Tel.: (022) 24927869, **Telefax:** 24915415, **E-mail: mumbai@newagepublishers.com**
- **New Delhi** 22, Golden House, Daryaganj, New Delhi-110 002, **Tel.:** (011) 23262368, 23262370
 Telefax: 43551305, **E-mail: sales@newagepublishers.com**

ISBN: 978-81-224-3146-9

C–16-08-9598

Printed in India at Glorious Printers, Delhi.
Typeset at Parallel Lines, E-8, Kalkaji, New Delhi.

NEW AGE INTERNATIONAL (P) LIMITED, PUBLISHERS
7/30 A, Daryaganj, New Delhi-110002
Visit us at **www.newagepublishers.com**
(CIN: U74899DL1966PTC004618)

Dedicated to

PROFESSOR JAYANT NARLIKAR

—our homage to a great teacher of physics and astrophysics

Cover photograph

The solar corona, photographed by the Skylab space station, during a total solar eclipse. The sun contains free electric charges in the form of electrons, protons and other particles. It has a strong magnetic field. The picture here shows streamers and loops formed mainly by high-speed electrons moving in the magnetic field. See Chapter 8 *(Courtesy: NASA/JPL)*

Preface

India has made several contributions to astronomy in the past. But there has been a lull over the last few centuries. We again find some research in modern astronomy from the late eighteenth century onwards in India, centred around observatories in Madras and Kodaikanal.

In the present century, a dozen institutions devoted to the study of modern astronomy and astrophysics have come up in the country. In fact, Meghnad Saha is considered the father of astrophysics throughout the world.

But unfortunately, not many bright students are coming forward to take up a career in astronomy and astrophysics in India. This is because of the fact that astronomy and astrophysics do not find a place in school and undergraduate curricula, unlike in the West.

The present venture is an attempt to motivate and inspire young students in schools and in the early years of university education, so that some of them may opt for a career in these areas. We sincerely hope that in the near future astronomy and astrophysics would find a place in undergraduate curricula. It would also cater to teachers and scientists not familiar with astrophysics.

This book would also be useful to the general public—engineers, doctors, lawyers, managers, and educated persons in other professions—having a certain background in science. Such people would find in this book a good deal of information about the solar system which would stimulate their thinking.

Astronomy is not just a classical subject. Its modern development depends on the new and sophisticated methods of observation and interpretation. We have tried to give an idea of how hi-tech methods have opened up several new vistas of the solar system, through the eyes of radio waves, infrared and ultraviolet rays, x-rays,gamma rays, cosmic rays, and neutrinos. India is making a meaningful contribution to several study programmes in these areas on the international scene.

We have tried to give in this book the most up-to-date and authentic information on planets, their satellites, asteroids, comets and other topics through numerous illustrations and tables. We have suggested simple demonstrations to understand basic concepts, such as the length of a day, the combined motion of the earth and the moon around the sun, the rotation of the moon about itself, and so on. Relatively difficult or allied concepts have been explained in several boxes spread throughout the book. About 50 photographs (colour and half-tones) from authentic sources and about an equal number of illustrations have been incorporated to make things vivid and clear.

PREFACE

New advances are taking place at a tremendous pace in the areas of astronomy and astrophysics. Some of the material presented in this book is so topical that it may become out-of-date in a few days or months. For example, the Magellan spacecraft which has been sent towards Venus (see Box 1.1) has gone into its orbit by the time this book appears in print, and would have sent us a good deal of information about the planet. At the same time, there are many things in this book which refer to more permanent aspects of the grand spectacle. Apart from presenting information, we have also tried to bring out the scientific method involved in the process of obtaining that information. We do hope this will help cultivate scientific temper among the readers.

AVINASH W JOSHI
NARAYAN C RANA

Acknowledgements

We are grateful to Professor Jayant V Narlikar, Inter-University Centre for Astronomy and Astrophysics, Pune University Campus, for discussing and going through parts of the manuscript, and for his encouragement. We are also indebted to Professor Arun S Nigavekar, Department of Physics, University of Poona, and Dr S Ananthakrishnan, Giant Metrewave Radio Telescope Project, Pune University Campus, for encouragement and timely help. We thank Dr. Arun D Tillu, Department of Physics, University of Poona, for discussions, and Professor K D Abhyankar, Professor Emeritus, Osmania University, Hyderabad, for critically going through many parts of the manuscript.

We are thankful to many agencies and individuals for providing photographs and allowing us to use them in this book. Their list appears separately.

We are highly indebted to the Department of Science and Technology, Government of India, for financial assistance which enabled us to write this book.

AVINASH W JOSHI
NARAYAN C RANA

Photo and Illustration Credits

The authors are grateful to the following individuals and agencies for supplying photographs/illustrations and giving permission to use them in this book. They are indicated by the following initials in the figure captions.

AVB:	Mr. Arun V. Bapat, Central Water and Power Research Station, Pune
EB:	Encyclopaedia Britannica, Chicago (U.S.A.)
CFC:	Mr. Charles F. Capen (U.S.A.)
JC:	Dr. Joe Churms, South African Astronomical Observatory, South Africa
JLE:	Professor J.L. Elliot, Department of Earth, Atmosphere and Planetary Sciences, Massachusetts Institute of Technology, Cambridge (U.S.A.)
GMRT:	Giant Metrewave Radio Telescope Project, Pune University Campus, Pune
HP:	Hansen Planetarium, Salt Lake City (U.S.A.)
IIA:	Indian Institute of Astrophysics, Bangalore
JPL:	Jet Propulsion Laboratory (U.S.A.)
LO:	Lick Observatory, Santa Cruz (U.S.A.)
NCRA:	National Centre for Radio Astrophysics, Pune University Campus, Pune
RLM:	Dr. R.L. Millis, Lowell Observatory, Flagstaff (U.S.A.)
NASA:	National Aeronautics and Space Administration (U.S.A.)
NRAO:	National Radio Astronomy Observatory, Charlottsville (U.S.A.)
PRL:	Physical Research Laboratory, Ahmedabad
SPAN:	*Span,* monthly magazine, the American Center, New Delhi
TG:	Professor Tom Gehrels, University of Arizona, Tuscon (U.S.A)
TIFR:	Tata Institute of Fundamental Research, Bombay
UPSO:	Uttar Pradesh State Observatory, Naini Tal
USO:	Udaipur Solar Observatory, Udaipur
LHW:	Dr. L.H. Wasserman, Lowell Observatory, Flagstaff (U.S.A.)

This book contains

25 Colour Plates
26 Black-and-White Photographs
50 Line Diagrams
18 Tables
29 Box Items

Greek letters used in this book

Ω [omega (capital)] — Longitude of the ascending node
ω [omega (lower case)] — Longitude of the perihelion from the ascending node
Υ [upsilon (capital)] — Vernal equinox
α [alpha (lower case)] — Symbol for a star
β [beta (lower case)] — Symbol for a star
H_α [H-alpha] — Symbol for the most prominent emission line in the hydrogen spectrum

Special units used

1 Å (angström) — 10^{-10} metre
1 μm (micrometer) — 10^{-6} metre
1 A.U. (astronomical unit) — 1.496×10^8 kilometres
1 light-year — Distance travelled by light in one year, 9.46×10^{12} kilometres
1 pc (parsec) — 3.26 light-years, 3.08×10^{13} kilometres

Contents

1

The Story Unfolds

Come one, come all! Here we are going to describe the story of the solar system. Read with diligence, and pay all the attention you can.

Did you say you already knew what the solar system was? The sun and a few planets and all that?

No, not really. That is the old story.

What about the satellites of all the planets, whose list grows longer every year, like the tail of *Hanuman*? What about the outer planets Uranus, Neptune and Pluto, which cannot be seen by the naked eye? Did you know that the last of these, Pluto, was discovered only in 1930? What if there are more planets yet unseen? Do you know some of the unfamiliar features of the familiar planets Mercury, Venus, the earth, Mars, Jupiter and Saturn? Did you know that the earth as a whole oscillates like a rubber football? Then there are millions of comets and satellites and asteroids. (Figure 1.1 shows the Halley's comet during its recent visit in 1985-86.) What about the sun and its wonderful features like the gases in it, sunspots, its interior, its magnetic field, and so forth?

Well, there are indeed several such aspects of the solar system, and many many more to be found in this book, which merit a closer look.

To tell you the truth, not even we know the full story—nor, for that matter, does anybody on this earth—for the story is really unfolding every day before our own eyes, like a grand, neverending suspense story. For example, about a hundred years ago, it was thought that Jupiter has four satellites (or moons) while Saturn has eight. These numbers rose to twelve and ten, respectively, about two decades ago. Now with closer observations based on pictures sent by the two Voyager spacecraft and other techniques, more satellites of many remote planets have been discovered. For Jupiter and Saturn, these numbers (of satellites) stood at 16 and 18, respectively, in 1990.

Voyager 2, during its passage near Neptune on August 24-25, 1989, discovered six more satellites of this planet, bringing the total number of its satellites to eight.

Similarly, the observations of comet Halley (Figure 1.1) during its recent appearance in 1985-86 have thrown much light on the nature of comets and have compelled scientists to revise their earlier estimates about their masses and structure.

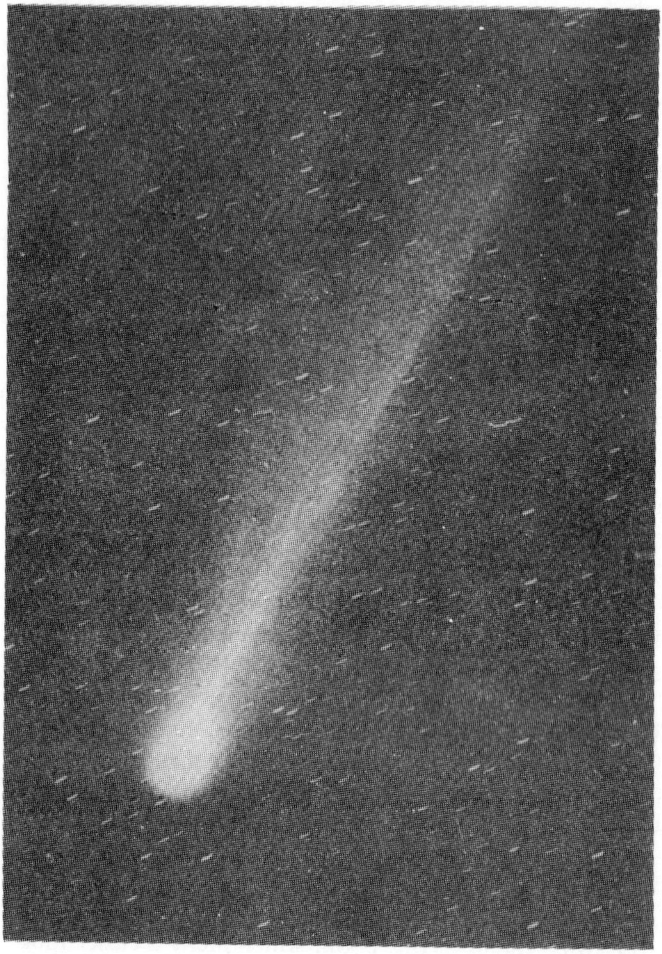

Figure 1.1 Comet Halley, during its recent appearance in 1985-86. (Photographed on March 19/20, 1986, with 51 cm Schmidt telescope at Kavalur; Exposure 30 minutes; *Courtesy:* IIA)

FUNDAMENTALS

Well, it would be a good idea to begin with the basics of the subject matter of this book. As it is known today, our solar system consists of the following components (see Figure 1.2):

(a) Its *prime star,* the sun;

(b) The *major planets,* which are Mercury, Venus, the earth, Mars, Jupiter, Saturn, Uranus, Neptune and Pluto. Nine of them are known at present;

(c) *Satellites* of all the planets;

(d) *The minor planets,* which consist of the asteroids, comets, etc.; and

(e) The space between all these objects, known as the *interplanetary space,* which itself consists of dirt, debris, meteroids, small particles, etc.

Box 1.1 *Launching spacecraft towards planets*

Since the first Soviet artificial satellite orbited the earth in 1957, the Soviet Union, the United States of America and European countries have sent spacecraft towards various planets, the moon and even the sun. (There have been about 72 astronomical missions until 1989.) This allows a closer study of the object concerned.

Apart from manned spacecraft which have landed on the moon, vehicles have been sent towards Venus, Mars and the sun, as well as the outer planets.

A space probe *Magellan* was launched from the US orbiting space shuttle *Atlantis* on May 5, 1989, towards Venus. It reached Venus in August 1990 after a 456-day trip in space. It will map about 90% of the surface of that planet, producing high resolution images which would be ten times sharper than any of the previous pictures.

A spacecraft has also been sent to Jupiter in October 1989. It will visit Venus and an asteroid on its way to Jupiter. For more detail, see Box 18.1.

The National Aeronautics and Space Administration (NASA) of the USA and the European Space Agency have undertaken a project, the Cassini project, to send a spacecraft towards Saturn in 1996. While the main module will go into an orbit around Saturn, another will take off from the main module towards one of its satellites, Titan. This module is expected to land on Titan with the help of a parachute. This project would no doubt bring Saturn and its satellites much closer to us.

The turn of the century will see a massive effort to launch unmanned and manned spacecraft towards Mars. NASA is likely to start these launches around the year 1998.

The relative sizes of the sun and the major planets are shown in Figure 1.3. Their 61 satellies are also shown, though not to scale. Some of the recently discovered satellites are so small that they would not show up in a diagram of this size.

In fact, many of these terms would be explained in this book as we go along with the story.

So, off we go.

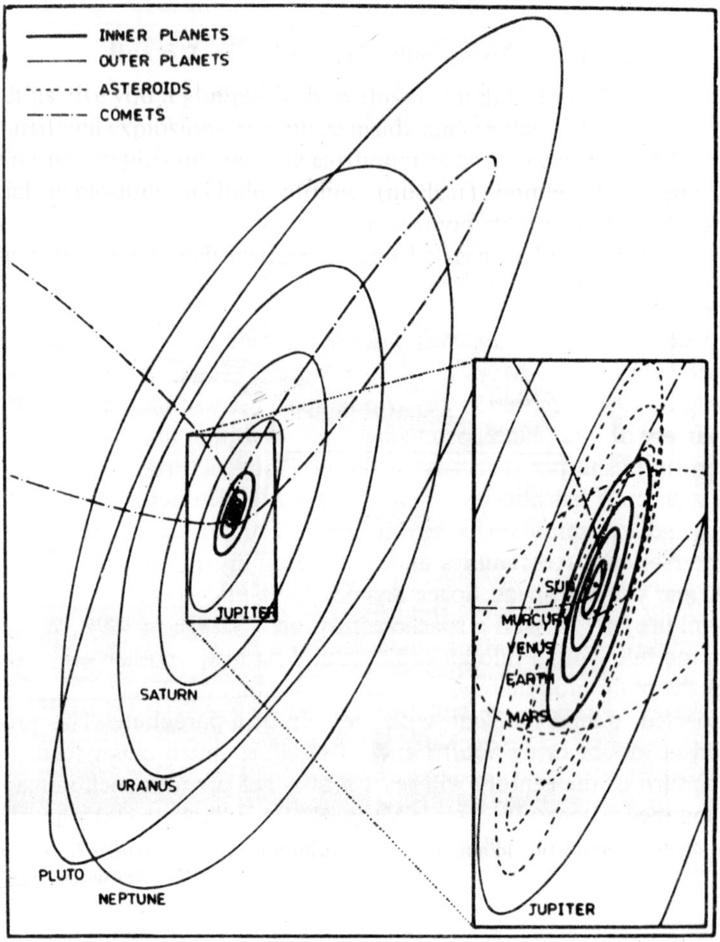

Figure 1.2 The contents of the solar system (not to scale)—planets, asteroids and comets. Orbits of nine planets and two comets are shown. Inset shows paths of some of the innermost planets and a few asteroids.

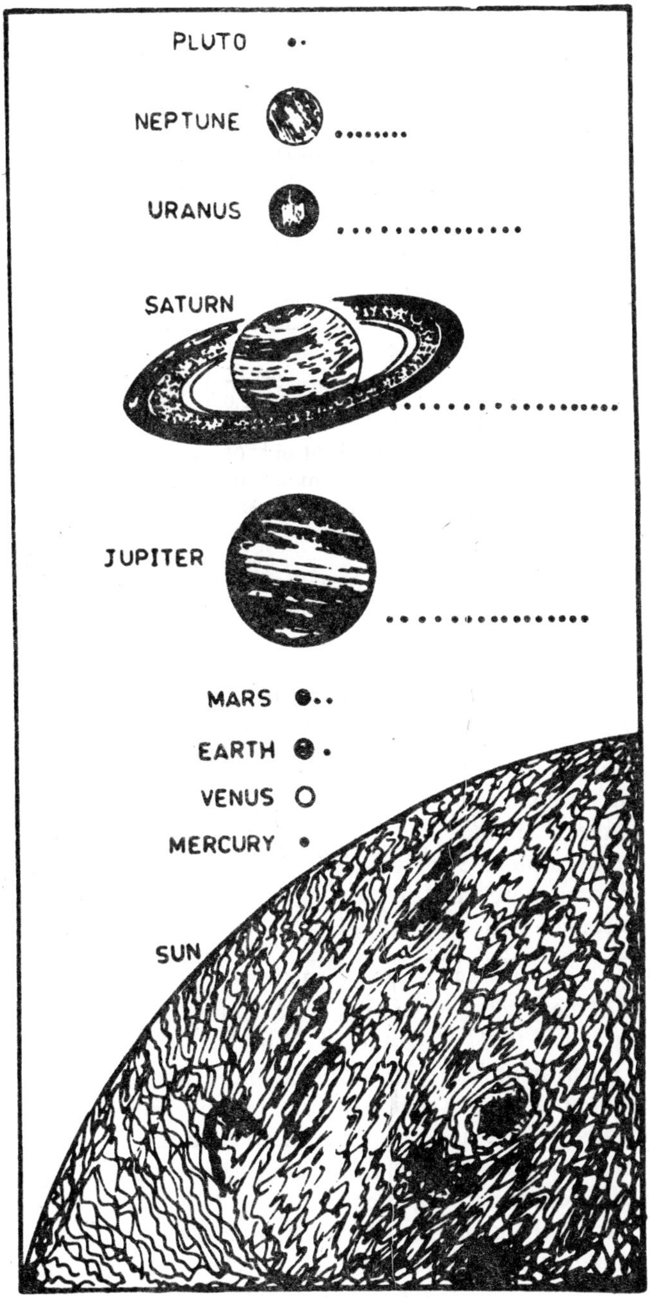

Figure 1.3 The relative sizes of the sun and the nine planets. Their 61 satellites are also shown, though not to scale.

2
Some Basic Definitions

Let us understand the various terms which we have used in the previous chapter in listing the contents of the solar system.

The most wonderful object in our solar system is the sun. It shines brilliantly and throws its radiations in all directions. A luminous body in the outer space which shines of its own is called a star (in Sanskrit, *Tārā*). There are billions of stars in the universe and you can see many of them shining in the sky on a dark night. Our sun is just one such star.

Stars are self-luminous. The sun, being a star, is naturally self-luminous too. All the objects in our solar system other than the sun are not self-luminous.

> **Box 2.1** *Luminous and non-luminous objects*
>
> An object which gives out visible radiation and shines is said to be a *luminous* object. Self-luminous objects are those which have their own fire or brilliance without any supply of energy from outside. For example, the flame of a lighted candle is a luminous object. But a wall, a table, a book and the moon are not luminous; we can see these objects only when light from some source falls on them. Stars are *self-luminous* because they shine and radiate their own energy.

There are some objects in the solar system which revolve around the sun. The earth is one of them. It revolves around the sun in a period which we call an *year*. Such an object is called a *planet* (in Sanskrit, *Graha*). Some other planets in our solar system are Mars, Jupiter, etc. which have been listed in the previous chapter under the title *major planets*.

You must learn to distinguish between a star and a planet. There are the following differences between them:

(a) A star is self-luminous, which means that it produces its own light and shines brilliantly. A planet is not self-luminous. A planet becomes visible to us because light from the sun falls on it, some of it is reflected from the surface of the planet, and enters our eyes.

(b) The stars (other than the sun) are extremely far away from us as compared to the planets. For this reason, the stars seem to be stationary in the sky. On the other hand, you can watch the motion of a planet with respect to the stars. We shall describe these motions in the next two chapters.

Figure 2.1 shows the motion of the planet Pluto, the outermost planet of the solar system yet known, in one day against the background of stars. This is how members of the solar system give themselves away when compared with stars.

(c) There is a simple observational method of distinguishing a planet from a star. Stars are so far away that they appear as point sources of light. Turbulence in our atomosphere causes them to twinkle. The planets, on the other hand, have visible discs made of many points. Each point in the disc twinkles, but the disc as a whole appears steady. Hence planets do not twinkle.

Then there are some objects which revolve around a planet. The moon is the best example of this. An object which revolves around a planet is called its *satellite* (*Upagraha* in Sanskrit). Thus the moon is a satellite of the earth. Many other planets also have satellites about which you will learn in this book.

There are innumerable small objects in the solar system which revolve around the sun in very elongated orbits. An object of this class is called a *comet* (*Dhūmaketu* in Sanskrit). Compared to these, the path of a planet around the sun is almost circular. This is illustrated in Figure 2.2.

There are millions of small bodies moving here and there throughout the solar system along different paths. *Asteroids* are those bodies which orbit the sun and whose orbits mainly lie between the orbits of Mars and Jupiter. Very often some small objects collide with the earth and are seen as *meteors*.

Fortunately for us, collisions of large bodies are extremely rare. One can only imagine what would happen if the earth were to collide with Mars! But the planets are separated from each other by large distances in space, and the stars are separated from each other by even larger, vast distances (see Chapter 19).

The entire space which contains the solar system is called the *interplanetary space*. Apart from the sun, all the planets, satellites and smaller bodies which cannot be seen even with a powerful telescope, it contains atoms, molecules, and various particles filling the whole space. All such small bodies are referred to as *dirt* or *debris,* and the smallest ones as solar wind and solar cosmic rays.

ASTRONOMY AND ASTROPHYSICS

We must also distinguish between *astronomy* and *astrophysics*. So far as the solar system is concerned, astronomy includes the study of the paths of the sun, the moon, the planets, the comets, etc. in space, and their positions with respect to the stars. It also includes the discovery of new planets, comets and satellites. On the other hand, astrophysics involves the study of the matter which these objects of the solar system are made of, how they must have been formed in the first place, their atmospheres, temperatures, pressures and other physical conditions. Of course, often there is appreciable overlap between the two. Originally astronomy was more of a mathematical science. Later it expanded its scope to include all the observational studies related to celestial bodies. Astrophysics has remained a physical science since its in-

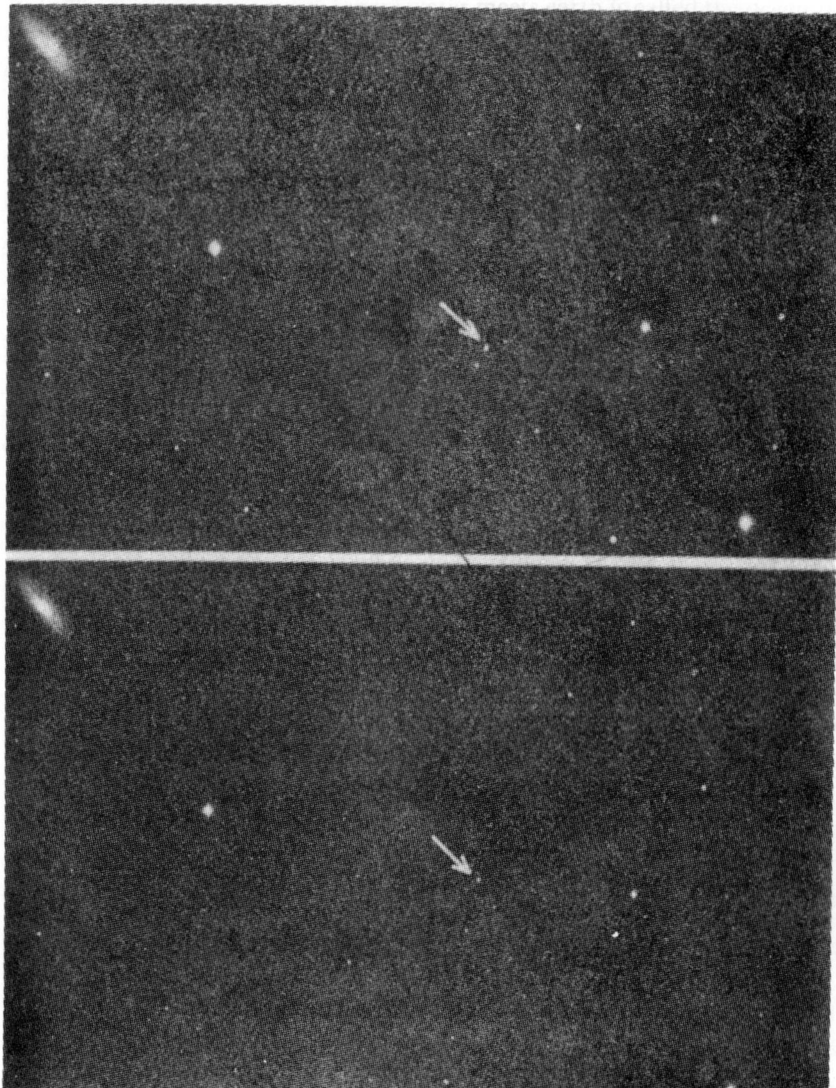

Figure 2.1 Two photographs showing the motion of Pluto in one day. The arrows indicate the position of Pluto. (Photographed with Crossley reflector. *Courtesy*: LO)

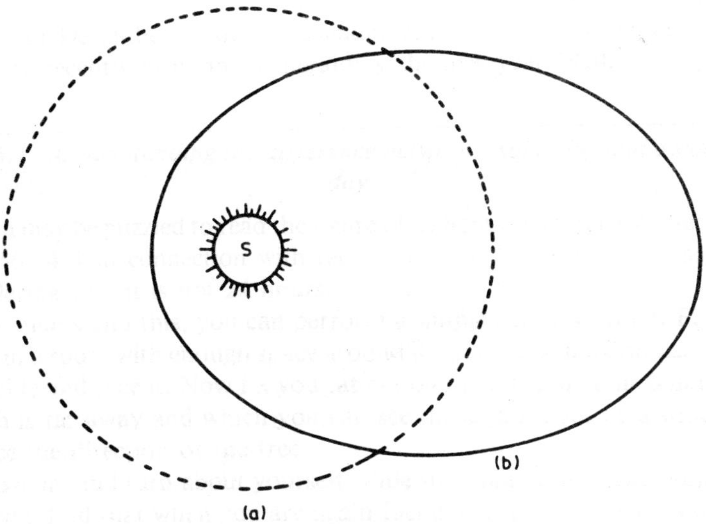

Figure 2.2 A sketch of the orbits of (a) planet, and (b) a comet around the sun.

ception in the early twenties of this century. They heavily rely on each other and are complementary.

You must bear in mind that both astronomy and astrophysics are *observational sciences.*

3
The Naked-Eye Solar System

Many of the phenomena occurring in the solar system can be seen by the naked eye, without recourse to any sophisticated instrument, although a lot more certainly remains invisible to the naked eye.

Thus we observe the sun, rising and setting daily. But if you notice carefully, the timing of sunrise and sunset shifts daily by a few tens of seconds. The days are longer in summer with the sunrise being early and the sunset late.

Choose two particular windows of your house to observe the sunrise and sunset throughout the year. Fix your attention on the tree or the building near which the sun appears to rise on a certain day. Do the same thing for the sunset. If you observe the sunrise and sunset every day, you will notice that it doesn't rise or set near the same tree or building every day. The point where it appears to rise or set shifts towards the north during the period 23rd December to 22nd June. This period is therefore called the *northerly motion* of the sun (*Uttarāyaṇa* in Sanskrit). During the remaining six months, that is the period 23rd June to 22nd December, these points move towards the south. This period is therefore called the *southerly motion* of the sun (*Dakshiṇāyana*).

Then, of course, you can observe the moon with the naked eye, with all its wonderful phases. The timings of the rising and setting of the moon and its phases have a very peculiar cycle. We will talk about this in more detail in later chapters.

You can observe solar and lunar eclipses with the naked eye. Every year, generally two to four eclipses (of the sun and the moon together) take place. During rare periods, as many as seven eclipses may take place in a year. (See Table 10.1 for the dates of solar and lunar eclipses which would occur up to AD 2000.) Figure 3.1 shows one such solar eclipse. You can learn a lot by observing these eclipses and the days on which they occur. For example, initially you are told that a solar eclipse may occur on a new moon day while a lunar eclipse may occur on a full moon day. But a solar eclipse does not occur on *every* new moon day nor does a lunar eclipse occur on *every* full moon day. You can find out the reason for this when we discuss the motion of the moon around the earth and of the earth around the sun in some detail later, in Chapter 10.

If you are diligent enough, you can observe the following planets of the solar system with the naked eye, when they are in the night sky: Mercury, Venus, Mars, Jupiter and Saturn. As you would see from Figure 1.2, the first two have their orbits inside the earth's orbit (the *inferior planets*) while

Figure 3.1 Solar eclipse; the sun's disc is completely hidden behind the moon. (*Courtesy:* IIA)

the other three lie beyond the earth's orbit (they are among the *superior planets*).

Mercury is the planet closest to the sun. In the sky also, it will therefore appear very close to it, always within 28° of the sun. Sometimes, it is so close to the sun or, in fact, 'on top' of it, that it cannot be seen due to the sun's brilliance. (See Figure 12.2 for a photograph of such an event seen through a telescope and Table 12.1 for future dates.) Even otherwise, it is very rarely observed, either a little before daybreak or a little after nightfall.

The next planet is Venus. In the sky, it lies within 47° of the sun. It can also therefore be observed only in the early morning (before sunrise) or in early hours of night (after sunset), though for a longer period (sometimes about three hours). It appears very bright and you can easily distinguish it, once you are introduced to it. Apart from the sun and the moon, it is the brightest object in the sky.

After Venus comes Mars. It can be seen as a red planet in the sky. It is commonly known as the *red planet*. Among its various names in Sanskrit, there are two names, *Lohit* and *Angāraka,* which refer to its redness. (See Colour Plates 9 and 10 for Mars.)

Then comes Jupiter. It is seen as a steady, fairly bright planet in the sky.

Finally, there is Saturn. To the naked eye, it is seen as a small but steady planet in the sky.

Develop the habit of observing the sun, the moon and the planets every day (or night). Of course, you may not be able to see all the planets at any particular time, except on very rare occasions (see Colour Plate 3 for such an event), since some of them may be on the other side of the earth. These would therefore not be visible to you.

There is much that can be learnt by observing the motion of the sun, the moon and the planets. We shall now discuss the motion of these objects in the sky. By these objects, we mean the sun, the moon and the planets.

4 The Motion of the Visible Planets

At this point, we must say that it is interesting to observe the planets in the sky. Watch their motion with time—over a few hours, a few days, and then over months and years.

Observe the motion of the sun or any planet in the sky. Observe the position of these objects against the 'fixed' stars in the sky. Make a note, preferably a written one, of the position of a planet relative to the sun and the stars. Notice the change in the position of the planet (a) in an hour or two (on the same night), (b) from day to day (at the same time every day), (c) over a period of several days or months, and (d) over a period of years. These observations will tell you a whole lot about the solar system.

THE DAILY MOTION

Let us first consider the gross motion of objects in the sky which hardly takes a few hours. This is part (a) of the observations mentioned above.

Consider, for example, a planet in the sky. When you observe it at a particular moment, suppose it appears to you to be situated in a certain group of stars. If you observe it for an hour or two, you will find that the planet has a motion towards the west *along with all the background stars*. What is really happening is that you, on the earth's surface, are moving towards the east because of the earth's daily rotation about itself. Thus the *diurnal motion* of a planet is a uniform westward motion in which its position with respect to the stars remains practically unchanged.

This is not surprising at all. Don't we observe that every planet or star, including the sun and the moon, rises in the east, moves from the east to the west in about 12 hours, and finally sets in the west? This is the apparent motion of all objects, which is seen because we on the earth are ourselves rotating around the earth's axis once a day from west to east. We cover an angle of 360° (that is, one full circle) in 24 hours, or 15° per hour. In fact, every object in the sky would possess this uniform westward motion.

Once again, we wish to make it clear that the diurnal motion of a planet is not its real motion with respect to the sun or the stars. It is only the apparent motion of a planet which occurs merely because the earth rotates about its own axis once a day.

To study the real motion of a planet, thus, you must forget its diurnal motion altogether. Concentrate only on its motion relative to the sun and the stars. Of course, its shift in the sky with respect to the stars over a day would be very small and unnoticeable (except for the moon). But if you observe it for a few days or months, you can easily notice its motion in the sky *with respect to the stars*.

The next part of the observation mentioned in (b) above is the day-to-day motion of an object. We have shown in Figure 2.1 the motion of Pluto in the sky in a day. Although every object in the solar system possesses this motion, you can most conspicuously observe it in the case of the moon. Let us therefore watch the moon on each night and notice its position relative to the sun and the stars.

THE MOON'S MOTION ACROSS THE STARS

The simplest thing to note is the separation in the sky between the sun and the moon. This will tell you how far behind, or ahead of, the sun the moon is. This can be done approximately by observing with the naked eye. But you can make a more accurate observation by noting the timings of sunrise and moonrise. These are given in all almanacs and occasionally even in newspapers.

Thus on a new moon day (when the moon is not visible to us because its dark side faces us), the moon must have risen and set with the sun. There is no appreciable angular separation between them.

The first day's moon after the new moon day is generally not visible because it is still very close to the sun in the sky. But the second day's moon after the new moon day can be seen for a few minutes after sunset. It sets about two hours after the sun does, and thus on this day the moon lags behind the sun by about two hours. (More correctly, this time lag is on an average one hour and 38 minutes on the second day.)

Box 4.1 *Angular separation between stellar objects*

The angle subtended by the rays coming from two objects at our eyes is called the *angular separation* between them. The angular separation between any two objects in the sky can lie between 0° and 180°. Two objects *appear* to be very close to each other when their angular separation is small, although in actual space they may be separated by a large distance. This is illustrated in Figure 4.1 (a).

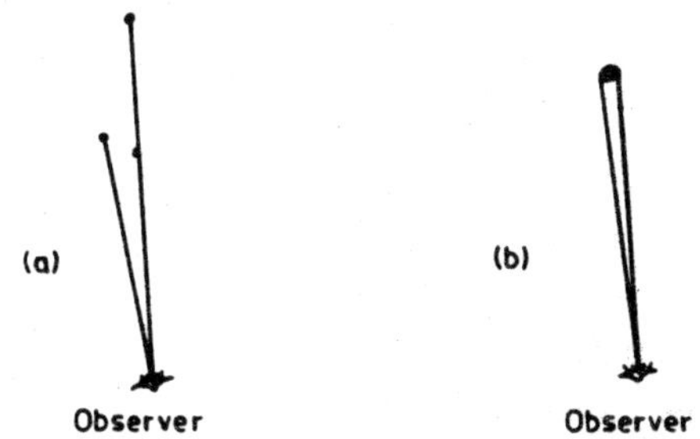

Figure 4.1 (a) The angular separation between two objects is the angle subtended by rays coming from them to our eye; and (b) The angular diameter of an object is the angle between the rays coming from its extremities.

The angular diameter of an object in the sky is similarly defined as the angle subtended by the two rays coming from its extremities at our eye. This is shown in Figure 4.1(b).

A full circle is divided into 360°, each degree is divided into 60 *minutes of arc* (or *arcmin*), and each minute is further divided into 60 *seconds of arc* (or *arcsec*). In symbols, for example, 10 minutes of arc are denoted by 10′ while 5 seconds of arc are denoted by 5″.

As an example, Pluto moves on an average through an angle of 14″.28 in a day (see Figure 2.1).

The first sighting of the moon after the new moon day is so important for the Moslem community that they break their fasts and offer special prayers on sighting the very first crescent moon of the month. Accordingly, every new month of the Islamic calendar begins on the actual sighting of the first crescent moon in the west.

Everyday thereafter, the moon lags behind the sun by a further 49 minutes and is thus visible in the sky for longer and longer periods. In terms of angle, the moon appears to move *eastward* relative to the sun by a little more than 12° every day. In about 15 days, the moon travels 180° in the sky and is diametrically opposite the sun. The earth is between the sun and the moon, and we have a full moon day. The moon rises about the same time as the sun sets and there is moonlight throughout the night.

This motion continues for the next 14 or 15 days until we have another new moon day.

The essential point to learn from these observations is that the moon moves towards the east in relation to the sun, covering slightly more than 12° of the sky every day.

MOTION OF PLANETS

Now take a look at some other planet, say Mars. Observe it for days and months and you will find its motion against the distant stars to be fairly irregular. Sometimes it moves towards the east and sometimes towards the west. This is illustrated in Figure 4.2.

In this figure, when the earth E is at positions 1, 2, 3, 4, 5, etc. the planet P is at the corresponding positions in its orbit. The lines 1-1, 2-2 etc. show the direction in which the planet would appear in the sky to an observer on the earth. During the period when the earth moves from position 1 to position 3, the planet P appears to be moving eastward against the stars. Between positions 3 and 4, the planet slows down and appears to be stationary for a few days. Then it reverses its motion and moves westward.

This phenomenon occurs because both the earth and Mars (or any other planet) revolve around the sun in approximately circular orbits. The period of revolution of Mars around the sun is about 687 days, which is about 1.88 years. This is 88% larger than the year on the earth. Because of this the earth overtakes Mars like a fast runner once in every 780 days. It is at this time that Mars appears to move backward towards west.

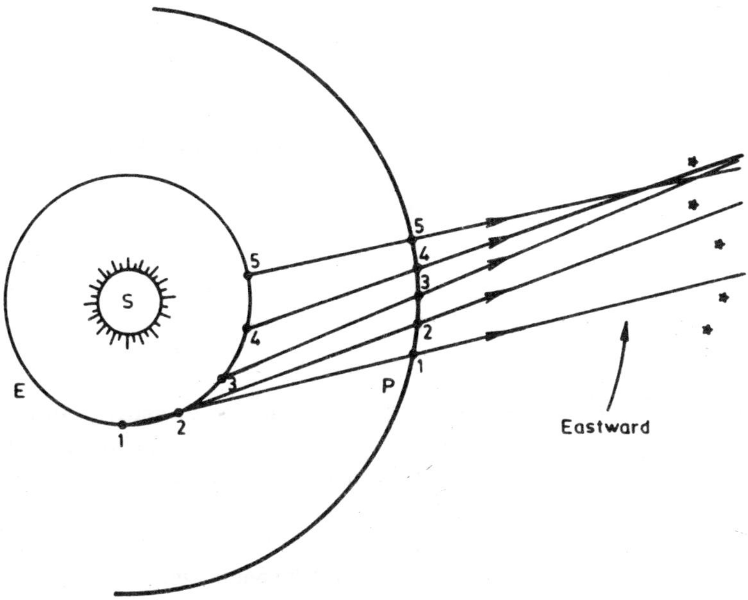

Figure 4.2 Forward and retrograde motions of a planet.

The motion of a planet in the sky towards the east is called its *direct* or *forward* motion (*Mārgī* in Sanskrit), while its movement towards the west is called *retrograde* motion (*Vakrī*). The moon does not have any retrograde motion because it is directly orbiting the earth. But Mars is not orbiting the earth; it orbits the sun, along with the earth itself. If we had an observation platform on the sun, the motion of Mars would appear to us to be fairly regular. From the earth, however, it appears to be erratic, a mixture of forward and retrograde motions.

We thus learn that the retrograde motion of planets is an apparent phenomenon which occurs because we are situated on the earth which is orbiting the sun, just as the uniform westward motion of all planets and stars is an apparent phenomenon caused by the rotation of the earth about itself.

In a similar manner, observe the planets Mercury and Venus whose orbits lie inside the orbit of the earth, and the planets Mars, Jupiter and Saturn whose orbits lie outside the orbit of the earth. For this, you will require some knowledge of the star groups in the sky against which we observe the planets. We shall postpone this to Chapter 7. We shall study our own earth in the next two chapters.

5

The Mother Earth

Colour Plate 1 is a view of the earth from the lunar orbit. The beautiful blue earth is hovering above the bleak and airless lunar horizon. Colour Plate 2 is a photograph of the earth taken by the crew of Apollo 17.

Let us recapitulate what you already know about our own planet, the earth. Perhaps you are aware of the fact that the earth is spherical in shape, like a large ball. It rotates about itself once a day (about 24 hours) and revolves around the sun once a year (about 365 days). It has an atmosphere surrounding it which extends to several kilometres in the sky.

ASTRONOMICAL AND PHYSICAL PARAMETERS

The earth is almost spherical in shape, with a mean radius of about 6370 km. It revolves around the sun along a path which is again almost circular. The distance of the earth from the sun is about 149.6 million kilometres (about 15 crore km). The mass of the earth is about 5.975×10^{24} kilograms, or approximately 6×10^{24} kg. (Place 24 zeros after 6 to get this number.)

Box 5.1 *Measuring masses of the sun and planets*

You would certainly wonder how the mass of a star or a planet is determined. You cannot weigh it in an ordinary balance nor can you hang it attached to a spring balance. Consider first the case of the earth.

The method lies in finding the distance of the moon from the earth and its period of revolution around the earth. Several persons recorded these observations very carefully during the fifteenth to the seventeenth century. This culminated in the ingenious and famous theory of universal gravitation proposed by Newton, who could then calculate the mass of the earth from his theory, provided the value of the universal constant of gravitation was supplied to him.

If one knows the orbital period of a satellite and the mean radius of its orbit around a planet, one can calculate the mass of the planet, since they are related to each other. This method does not work for a planet which has no satellite.

The sun has several planets going around it. The mass of the sun has been calculated by knowing their distances from the sun and their periods of revolution. But again, it cannot be used for any other star because we have not been able to observe a planet of any other star directly. Other methods have to be used in such cases.

Look at Figure 5.1 to understand the various terms we will use in this book. This figure shows the earth in its path around the sun. This path lies in a plane known as the *plane of the ecliptic.* The term *ecliptic* arises from the fact noticed by our ancients that when eclipses of the sun and the moon occur, both of them invariably lie on this plane.

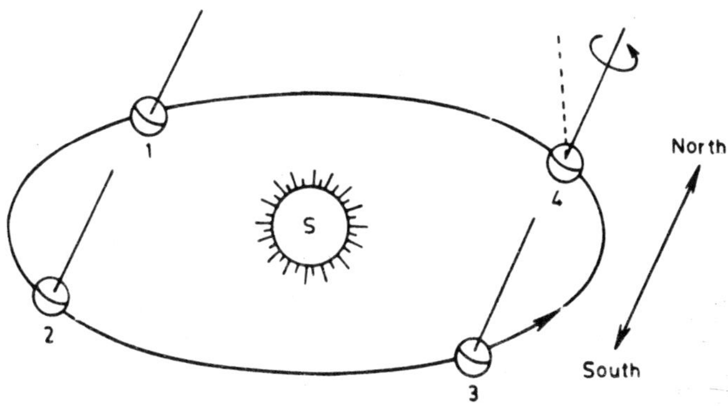

Figure 5.1 Position of the earth in its orbit around the sun at different times of the year. The earth is at Point 1 on about March 21, at Point 2 on June 22, at Point 3 on September 23, and at Point 4 on December 23.

The axis on which the earth rotates about itself is inclined to the line perpendicular to the ecliptic through an angle of about 23°.45. The points where this axis cuts through the earth's surface are known as the north and the south poles. For reasons which we shall explain later (see Chapter 6), these are called *geographic* north and south poles. You can thus see that at both the poles, there will be sunshine for six months and darkness for the other six months. So from March 21 to September 23, there will be a six-month long day at the north pole and an equally long night at the south pole. The situation will be reversed during the remaining six months, that is, September 23 to March 21. During this time, the north pole would have a long night and the south pole would soak in brightness. This perspective would be clear from Figure 5.1.

If we had an observation platform somewhere above the north pole and were looking *down* at the earth, it would appear to be rotating around itself in an anticlockwise manner. It also revolves around the sun in the same sense.

The imaginary circle on the earth's surface which is perpendicular to the axis of rotation and which cuts the earth into two equal halves (like an orange cut into two halves) is known as the *equator.* Both the north and the south poles are equidistant from the equator. The half of the earth on one side of the equator containing the north pole is called the *northern hemisphere* while the other half containing the south pole is called the *southern hemisphere.*

Approximately on March 21 and on September 23, the sun's rays at noon are exactly at 90° above the horizon at the equator. These dates and the corresponding points on the celestial sphere are called the *equinoxes* (also see

Chapter 7) On these days, the sun rises at 6 am and sets at 6 pm (local time) at *all* points of the earth (except the poles); hence the name 'equinoxes', meaning equal nights. The sun can be seen from both the poles on these days.

March 21 is known as the *vernal equinox* while September 23 as the *autumnal equinox* (*Vasantasampāta* and *Sharadasampāta* in Sanskrit, respectively). Vernal equinox is the first day of spring and autumnal equinox the first day of autumn in the northern hemisphere. The situation is exactly the opposite in the southern hemisphere.

SEASONS

The inclination of the earth's rotation axis to the plane of the ecliptic is the cause of the seasons. During the six months from March 21 to September 23, it is summer in the northern hemisphere and winter in the southern hemisphere. During the remaining half of the year the situation is reversed, so that it is summer in the southern hemisphere and winter in the northern hemisphere. As you go from the equator to either pole, you will see the phenomenon of longer days in the summer and longer nights in the winter. Near either pole the day lasts for six months followed by a six-month-long night.

ORIGIN OF THE EARTH

It is believed that the earth and the other planets were formed along with the sun from a large cloud of dust and gas by gravitational contraction. The age of the earth is somewhere between 4.5 billion and 4.6 billion years. (One billion is a thousand million or a hundred crores.) The best estimate of the age according to the present theories is 4.55 billion years. This means that before 4.55 billion years, there was only cosmic gas and debris of rocks which then condensed to form the solid and liquid parts of the earth.

THE EARTH'S SURFACE

There is a large store of water on the earth's surface in the form of oceans. Water covers 71 per cent of the earth's surface, leaving only 29 per cent for the land.

The average annual surface temperature on the earth varies from 32°C to −32°C. But the highest recorded temperature (in shade) has been 58°C (in the Libyan desert) while the lowest recorded temperature is −87°C (Antarctica).

The highest peak on the earth is *Mount Everest* in the Himalayan range. Its height is a little over 8 kilometres above the mean sea level.

INSIDE THE EARTH

How does the inside of the earth look like? What is it made of? How does one get information about the innermost depths of the earth?

There has been much research during the last two decades to answer the above questions. Scientists are now using very sophisticated new techniques and making extremely delicate measurements. This has given us information about regions where no person can reach.

The entire sphere of the earth consists of three main regions: *crust, mantle* and *core.* The core further consists of a solid *inner core* and a liquid *outer core.* If the earth's radius is taken as 100 units, the radius of the inner core would be about 20 units, that of the outer core 54 units, while the thickness of the crust would only be 1.5 units. These are shown in Figure 5.2.

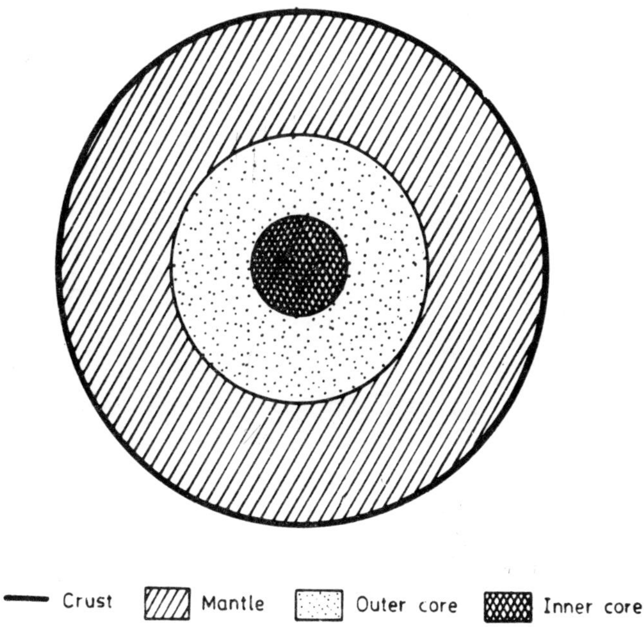

Figure 5.2 The inside of the earth; crust, mantle, outer core and inner core.

As one starts from the centre of the earth, the innermost spherical region with a radius of about 3480 km is called the *core.* On the other hand, the topmost layer having a thickness of about 70 to 100 km is called the *crust.* The middle region between the two is the *mantle.* This division has been made on the basis of the contents and the density of matter in these regions. Most of the interior of the earth is in the solid form while some part—particularly the outer part of the core—is found to be in a liquid state, indicating high temperatures. The temperature in the region of the core of the earth is about 5000°C to 7000°C.

The interior of the earth is not yet fully understood and there is a vast scope for research in this area.

Box 5.2 *Exploring the interior of the earth*

Let us give you a glimpse of how this research is done. Several natural and artificial explosions take place inside and on the surface of the earth. The natural explosions include earthquakes and volcanoes, while the artificial explosions include atomic (nuclear) bombs. Scientists have developed an instrument known as the *seismometer* to record and measure these events. One such instrument can be seen in Figure 5.3.

Figure 5.3 A seismometer, used for measuring the motion of parts of the earth during an explosion. (*Courtesy:* AVB)

When an explosion occurs, there is a wave of vibrations which passes throughout the earth. This wave can be picked up by a seismometer in any other part of the earth. Presently the entire surface of the earth is covered with more than 2000 such instruments. This is how an earthquake in Mexico or China or an atomic explosion in the United States of America can be recorded in, say, India. In addition, the passage of such a wave through the earth gives much information about the nature of the inner parts of the earth.

THE ATMOSPHERE

But for the earth's atmosphere, the sky would appear dark and all the stars would be visible to us even during the day, along with the sun. Fortunate are the astronauts who get this view when their spacecraft goes beyond the earth's atmosphere.

The earth has an atmosphere covering it like a blanket all around it. It consists of various gases and extends to a height of about 800 kilometres.

As you may already know, the main constituents of the atmosphere are nitrogen and oxygen. There are other trace elements which are also important. Table 5.1 lists all the constituents of the atmosphere along with their abundances. The atmosphere is very useful to us because it stops the intense high energy radiations coming from the sun. Without this atmosphere, it would have been impossible for life to develop on the surface of the earth.

Table 5.1 The major constituents of the atmosphere

Constituent	*Percentage (by volume)*
Nitrogen	78.084
Oxygen	20.946
Argon	0.934
Carbon dioxide	0.031
Ozone	10^{-4} to 10^{-3} in stratosphere (ozone layer) 10^{-6} in troposphere
Neon	0.00182
Water vapour	highly variable, responsible for clouds, practically nil in stratosphere and above
Other elements	helium, krypton, methane, hydrogen, carbon dioxide, oxides of nitrogen; each less than 0.0005

Because of the gravitational attraction of the earth, the density of the atmosphere is the largest near its surface. The gases become rarer as we go to greater heights. Finally, as we go beyond 800 km, the density of the atmosphere becomes negligible.

On the other hand, there is solar radiation falling on the top layers of the atmosphere. Only a part of it penetrates down to the earth. The solar radiation from the top and the earth's gravitational pull from down below give rise to various layers of the atmosphere with different properties. The layers of the atmosphere and their average heights are listed in Table 5.2. The actual heights of these layers vary from hour to hour during the day.

Table 5.2 Different layers of the atmosphere

Name of layer	*Range of height (in km)*	
	Minimum	*Maximum*
Troposphere	0	10–15
Stratosphere	10–15	50–55
Mesosphere	50–55	80
Thermosphere	80	480
Exosphere	480	about 800

The temperature of the atmosphere shows an interesting behaviour. It falls initially to $-55°C$ up to a height of 20 km, then rises to 10°C at a height of about 50 km. It falls again to a very low value of $-120°C$ at about 90 km and then rises continuously with height. Figure 5.4 shows the variation of temperature of the atmosphere with height above the surface of the earth.

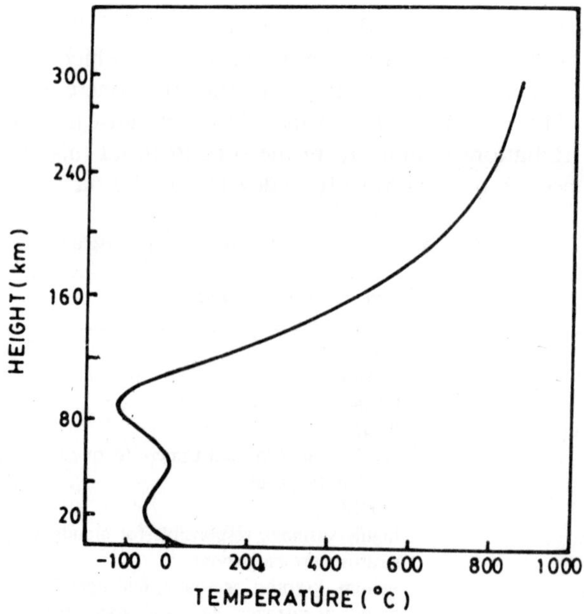

Figure 5.4 The temperature in the various layers of the atmosphere. (Based on *COSPAR International Reference Atmosphere,* North-Holland, Amsterdam, 1965.)

6

Some Unfamiliar Features of the Earth

MAGNETIC FIELD

The earth has several interesting and normally unfamiliar features. For one thing, it has a magnetic field all round it which extends to a height of several tens of thousand kilometres above the earth's surface. Its axis lies in an approximately north-south direction. The magnetic field makes an angle with the surface of the earth which changes from place to place.

It was noticed by scientists long ago that a magnetic needle or a freely suspended magnet always comes to rest in the north-south direction. In fact, this is how the earth's magnetic field was discovered.

Now it is believed that this magnetic field is created because of the presence of huge amounts of molten nickel and iron in the interior of the earth. (Remember that we use iron to produce magnets.) It is found that the earth behaves *like* a huge bar magnet, with its *magnetic* north pole near the geographic south pole and vice versa. There is further evidence that the earth's magnetic field has changed and even flipped direction in the past.

You must note that the magnetic north and south poles are quite distinct from the geographic north and south poles. In fact, the magnetic north pole is somewhere near the geographic south pole of the earth, and vice versa. The magnetic north pole is roughly situated at 66°.5 south latitude and 140°.5 east longitude, while the magnetic south pole is at 75°.5 north latitude and 100° west longitude.

The earth receives different kinds of high energy radiation not only from the sun but also from the entire cosmos. Much of this radiation consists of electrically charged particles. When these particles come near the earth, they encounter its magnetic field. The magnetic field bends the path of these electrically charged particles and produces several spectacular phenomena. One such phenomenon is known as *aurora* which is the name given to a brilliant coloured light seen in the sky in the arctic and the antarctic regions (approximately beyond 60° north as well as south latitudes).

OCEANS

The oceans are indeed a magnificent feature of the earth. They play a great role in controlling the temperature and climate.

They have huge currents, of warm as well as cold water. An ocean current is like a large river flowing *through* the ocean. These ocean currents are very useful for navigation.

The bottom of an ocean where water meets earth is called the *ocean floor*. The ocean floor contains many ridges, volcanoes and mountains. Figure 6.1 shows what a mountain on an ocean floor may look like. The deepest ocean occurs at a place called *Challenger Deep* near Mariana Islands in the Pacific Ocean.

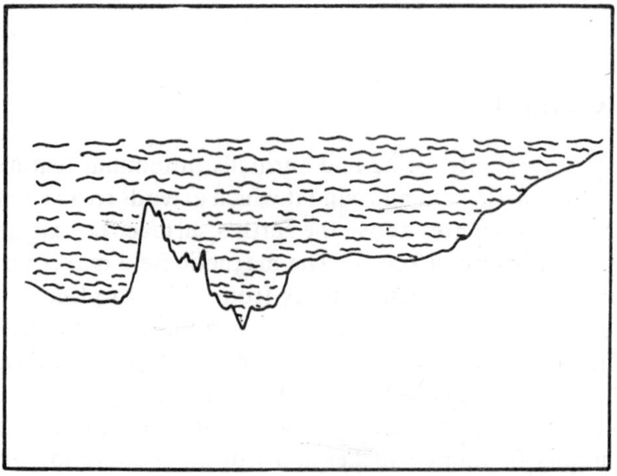

Figure 6.1 An ocean floor with hills and mountains looks like this.

You can also see the phenomenon of *tides* in the sea. The water level of the sea rises and falls twice a day. When the water level rises, it is called *high tide,* and when it falls, it is called *low tide.* Not all high tides are equally high. During the new moon and full moon days, the high tides assume the highest levels, called *spring tides,* and the low tides the lowest levels, called *neap tides.*

The gravitational attraction of the moon and the sun is responsible for raising the tides.

EARTH'S ROTATION

You will be surprised to know that the shape of the earth is not perfectly spherical, as an ideal table-tennis ball should be. It is slightly elongated, though not as much as an egg. It has a somewhat larger bulge at the equator than at the poles. The difference is about 21.4 km (in a mean radius of about 6370 km, which is about 0.335%). The earth assumes this shape due to its rotation around itself about an axis passing through its centre.

The period of rotation of the earth about its own axis with respect to the fixed stars is 23 hours 56 minutes 4.1 seconds. Some variation at the level of milliseconds has been seen in this period with modern sophisticated and delicate instruments.

The earth's rotation about itself is currently *slowing down,* though at an extremely slow rate. This means that the period of earth's rotation is increasing.

That is why all the clocks on the earth are stopped for a second at midnight of June 30 or December 31 of the calendar year as and when required. This was done recently at midnight heralding the new year 1990.

Box 6.1 *Understanding the difference between a solar day and a sidereal day*

You may be puzzled to read the figure of 23 hours 56 minutes 4.1 seconds ($23^h\ 56^m\ 4^s.1$ in connection with the rotation of the earth. You may be wondering why it is not 24 hours.

To understand this, you can perform a simple demonstration. Place a table in a room with enough space around it. Stand at some distance from the table and face it. Now fix your attention on a tree or some landmark which is far away and which you can see through a door or a window. Notice the direction of the tree.

Take one full turn about yourself while standing at the same position. You will find that when you are again facing the table, the tree is in the same direction as before.

Now take a slow turn about yourself in an anticlockwise manner and move around the table in *the same manner,* that is, in the anticlockwise manner. It is like a merry-go-round in which each cradle turns around itself and all the cradles revolve on a horizontal wheel.

You will notice that the position when the tree is again in the same direction is not the same when you are facing the table. This is illustrated in Figure 6.2. Suppose you start performing the two motions at the point A marked in Figure 6.2. When you concentrate on the tree and forget the table altogether, you will see that you have actually completed one rotation about yourself at the point B. At this point B, the tree appears to you in the same direction as from point A. But you are not facing the table until you move to the point C. So far as the distant tree is concerned, you have completed one rotation about yourself at B. But the table appears in the direction as before only when you move to C, which takes a slightly longer time.

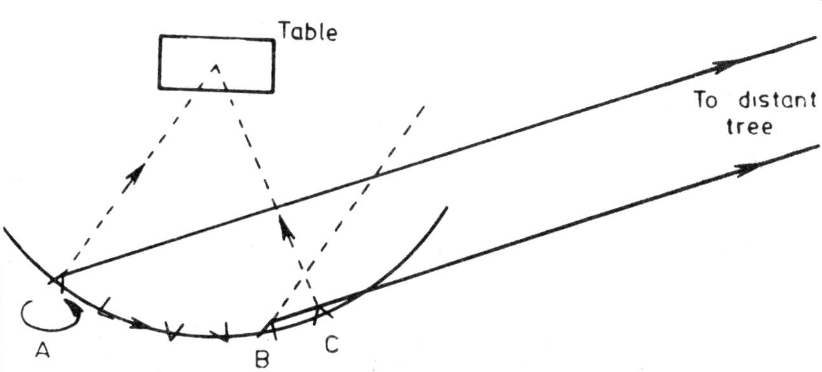

Figure 6.2 A boy taking a turn about himself and at the same time going around a table.

One important point to learn from this demonstration is this: *If you are alone somewhere in space with nothing else in sight and you start turning about yourself, you will have* no way of knowing *when you have completed one rotation.* You can try this with your eyes closed. This means that we can count rotations only with respect to some other object.

You should remember that the earth rotates about its axis and also revolves around the sun. The two motions take place simultaneously. Thus when the earth completes *one* rotation about itself with respect to the stars, it is not at the same place in its orbit around the sun where it was the previous day.

Replace the table in the demonstration in Box 6.1 by the sun, yourself by the earth, and the tree by a star. This motion is shown in Figure 6.3. You

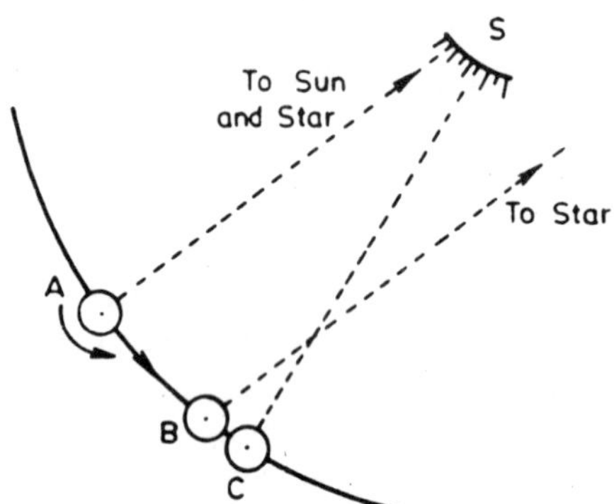

Figure 6.3 Rotation of the earth about itself and its revolution around the sun.

will now have no difficulty in understanding that the earth rotates about itself with respect to the stars in a slightly smaller period of time than with respect to the sun. As the earth completes one rotation around itself, it goes from point A to point B (Figure 6.3) along its path of revolution around the sun. All the stars (except the sun) would appear to be in the same direction from B as from A. But it takes a little longer (to point C) to get the sun in the same relative direction as form A.

We define a *solar day* as the duration between two moments when the sun is at the zenith. We have arbitrarily divided this duration into 24 hours. Therefore one solar day contains 24 hours *by definition.* But with respect to the stars, it takes $23^h\ 56^m\ 4^s.1$ for one rotation—after this time, a star (except the sun) will appear in the same direction as before. This is called a *sidereal day.* For this reason, the stars appear to rise and set 3 minutes 55.9 seconds earlier each day.

The axis around which the earth rotates about itself is not perpendicular to the plane of the ecliptic; it is inclined to the latter roughly by 66°.5. We have already told you that this inclination of the axis is the cause of the seasons on the earth. But even this axis does not point in a fixed direction. It rotates in a circle of angular radius 23°.5 once in 25 800 years, like the axis of a rotating top precessing about the vertical. Presently the axis points in a direction about 1° away from the star Polaris (*Dhruva Tārā* in Sanskrit) which is why Polaris always appears to us in the same direction. 12 900 years from now, the earth's axis would point in a direction making an angle of 47° (23°.5 twos are 47°) from Polaris, approximately in the direction of the star Vega. The perspective will be clear from Figure 6.4. This phenomenon is called the *precession of the earth's axis.*

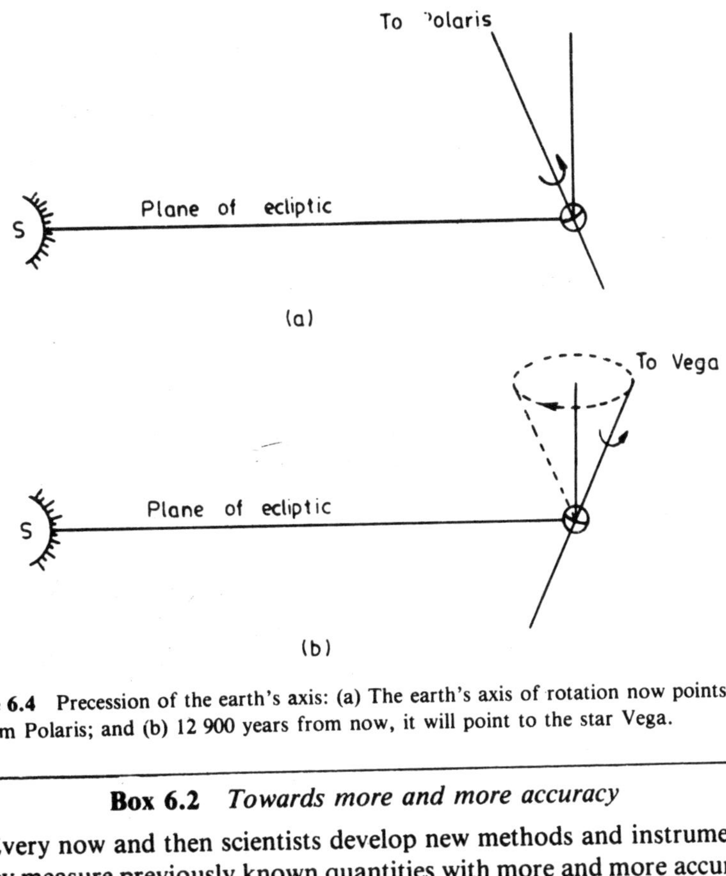

Figure 6.4 Precession of the earth's axis: (a) The earth's axis of rotation now points about 1° from Polaris; and (b) 12 900 years from now, it will point to the star Vega.

Box 6.2 *Towards more and more accuracy*

Every now and then scientists develop new methods and instruments. They measure previously known quantities with more and more accuracy by means of these. Consider the quantities such as the radius of the earth, its distance from the sun, the period of rotation about itself and revolution around the sun, the distance of the moon from the earth, etc. These quantities have been measured for several centuries and are still being measured with increasing accuracy. The present methods give us not only better values but also indicate any small variation that may occur in these parameters due to subtle physical effects.

For example, the average sun-earth distance is now accurately known to be 149 597 871 km.

ASTRONOMICAL PARAMETERS

The average distance (averaged over time) of the earth from the sun is 149 597 871 km (about 150 million or 15 crore km). This distance is taken as the unit for measuring distances of objects in the solar system. It is known as the *astronomical unit* (AU). It is very convenient to measure distances of objects in the solar system in terms of this unit.

The orbit of the earth around the sun is not exactly circular. It is like a slightly elongated circle, known as an ellipse. (See Chapters 11 and 12 for details of planetary orbits.) For this reason, the distance of the earth from the sun varies with the time of the year. The earth is nearest to the sun between January 2 and 4, when the distance between them is 147 million km. It is farthest from it between July 2 and 4. The distance between them in this position is about 152 million km. The difference between the two distances is slightly more than 5 million (50 lakh) km (which is about 3.3%).

The period of revolution of the earth around the sun is 365 days 6 hours 9 minutes (or $365^d\ 6^h\ 9^m$). The earth moves in its orbit around the sun with a mean speed of 29.8 kilometres per second or 1.073 million kilometres per hour.

DYNAMIC INTERIOR OF THE EARTH

Recent studies about the inside of the earth have further revealed some unimaginable phenomena. The earth is not a dumb, static ball but a dynamic engine powered by heat. The earth shudders, it vibrates, it shakes, and it also oscillates (expands and contracts) like bellows! Such oscillations, known as *football oscillations,* were discovered for the first time during the study of earthquakes in Chile in May 1960 and in Alaska in March 1964. They were seen to have time periods ranging from three minutes to about an hour. This field of study has revealed many bewildering phenomena since then.

7

A Bit of Star-Gazing

You must have noticed by now that all the planets as well as the sun and the moon appear to move in the sky in a belt of a small width stretching around the earth roughly in the east-west direction. This belt is known as the *zodiac*. There are several star-groups spread all over the sky. We will introduce to you only those star-groups which lie on the zodiac. Let us get a little more familiar with the major star-groups in this belt.

The stars *appear* to be stationary because they are far away from the earth. Actually, they are also moving, some with tremendous velocities. But their motion in the sky is almost imperceptible in the life-span of a person or even in a few centuries. Since we are interested in the solar system here, we would regard the stars as *fixed* and only consider those aspects which are necessary for a better understanding of the motion of planets.

The entire visible sky in the belt in which the sun, the moon and the planets apear to move has been divided into twelve equal parts. Thus each part covers 30° of the zodiac in the sky. Each part is called a *zodiacal division,* which is recognised by the prominent stars in that region. The star-groups are called *zodiacal constellations* (*Rāshi* in Sanskrit). The twelve constellations do not divide the zodiac into equal parts, like the zodiacal divisions, although they are approximately equal. It may be pointed out that the zodiacal signs used in newspaper columns are in some way related to the zodiacal constellations. They are called 'zodiacal' as our ancients imagined some animals depicted on the canvas of the sky.

Apart from the twelve constellations in the zodiac, there are, of course, other constellations in the entire sky. But in this book, we would mainly refer only to the twelve zodiacal constellations (except in Chapter 19).

ASTRONOMICAL COORDINATE SYSTEM

We need a coordinate system to specify the position of a celestial object (a planet, a comet or a star) in the sky. We must have two coordinates to find out the direction in which to look for an object and one coordinate to specify its distance from the earth.

For a moment, let us forget that celestial objects are at different distances from the earth. We consider all such objects to be moving on an imaginary *celestial sphere.* We now want two coordinates to specify the direction of an object on this sphere with respect to the earth.

This is exactly like specifying the position of a place on the spherical surface of the earth by giving its *latitude* and *longitude*. We take the zero of latitude at the equator and count up to 90° N to the north pole and up to 90° S to the south pole. Similarly we fix the zero of longitude at the meridian passing through the town of Greenwich and count up to 360° eastward.

In the astronomical coordinate system, the coordinate equivalent to latitude is called *declination* (Dec for short), while the coordinate equivalent to longitude is called the *right ascension* (RA for short). Figure 7.1 shows the celestial sphere with the earth at the centre.

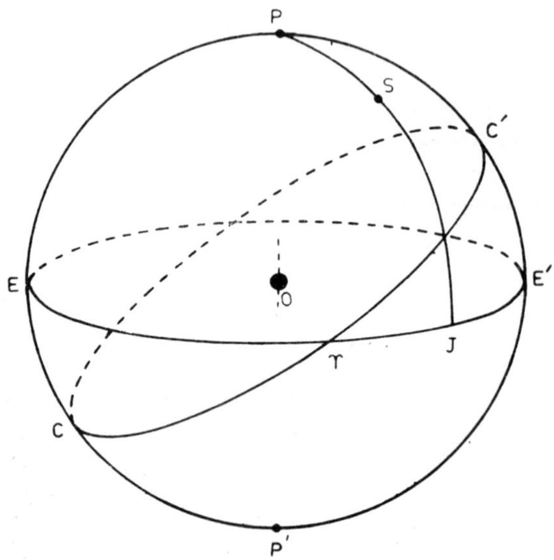

Figure 7.1 The celestial sphere for an observer on the earth, which is assumed to be at the centre O.

If the axis of the earth's rotation is extended in both the directions, the imaginary points P and P′ at which it would intersect the celestial sphere are called the *celestial poles;* see Figure 7.1. The imaginary circle E♈E′ at which the the earth's equatorial plane intersects the celestial sphere is called the *celestial equator.* The imaginary circle C♈C′ on the celestial sphere along which the sun *appears* to move in its annual sojourn of the stars is called the *ecliptic.* The points of intersection of the celestial equator and the ecliptic are called the *equinoxes.*

Although the celestial poles and the equinoxes are imaginary points, their directions are definite—they are physical quantities and can be specified by measurable parameters.

The celestial equator is taken as the zero of declination. Declination is counted from 0° to 90° in the northern hemisphere and 0° to −90° in the southern hemisphere of the celestial sphere.

The declination of the celestial north pole is thus 90° while that of the celestial south pole is −90°. The declination of the sun on March 21 and

September 23 is 0°. The sun's apparent path, that is the ecliptic, crosses the celestial equator on these days. In fact, the sun passes through the equinoxes on the above two days. On these days, it appears closest to zenith at noon from any point on the equator. The sun's declination varies between a maximum of 23°.5 (on June 22, the *summer solstice*) and a minimum of −23°.5 (on December 23, the *winter solstice*).

The equinox at which the sun passes from the southern hemisphere to the northern hemisphere is called the *vernal equinox* (VE), while the other one at which it passes from the northern hemisphere to the southern hemisphere is called the *autumnal equinox*. These events occur roughly on or about March 21 and September 23, respectively, every year. In Figure 7.1 ♈ is the VE while the autumnal equinox is diametrically opposite.

The vernal equinox is taken as the zero of the right ascension. This coordinate is measured positive eastward, which is the direction of the sun's apparent annual motion among the stars.

One could have measured the right ascension from 0° to 360° (to make a full circle). For convenience, however, this is converted into an equivalent of 24 hours, to match with the period of the apparent diurnal motion of the sun around the earth. Thus 15° are equivalent to 1 hour.

A degree as well as an hour is further subdivided into minutes and seconds. Thus an angle of 15′ (15 arcmin) would correspond to 1^m on the time scale. Conversely, 1^s on the time scale would correspond to an angle of 15″ (15 arcsec).

The RA of the sun when it is at VE is zero, at the summer solstice it is 6^h (90°), at the autumnal equinox it is 12^h (180°), and at the winter solstice it is 18^h (270°).

Consider a star which appears on the celestial sphere at S (see Figure 7.1). Let PSJ be the arc of the great circle passing through the north celestial pole and the star, intersecting the equator at J. Then the angle subtended by the arc SJ at the earth is the Dec of the star while ♈J is its RA.

ZODIACAL CONSTELLATIONS

The conventional and the modern English names and the corresponding Sanskrit names of the 12 signs of the zodiac (or constellations) are given in Table 7.1. The modern English name of a constellation corresponds to the modern meaning of the ancient Latin word. The prominent star-groups in the 12 zodiacs are shown in Figure 7.3 relative to the sun, the earth and ecliptic. Figure 7.4 shows each of these constellations in more detail.

Now you can go outdoors on a starlit night and identify, with the help of Figure 7.3 or Figure 7.4, some of the constellations. Remember that all of them will lie along a somewhat tilted belt in the east-west direction, that is, the zodiac.

Of course, at any one moment, you will be able to observe at the most six constellations in the sky because the other six will lie hidden from your

Table 7.1 The twelve zodiacal constellations (Rāshi)

	Conventional English name	*Modern English name*	*Sanskrit name*
1	Aries	Ram	*Mesha*
2.	Taurus	Bull	*Vrishabha*
3.	Gemini	Twins	*Mithuna*
4.	Cancer	Crab	*Karka*
5.	Leo	Lion	*Simha*
6.	Virgo	Maiden	*Kanyā*
7.	Libra	Balance	*Tulā*
8.	Scorpio	Scorpion	*Vrishchika*
9.	Sagittarius	Archer	*Dhanu*
10.	Capricorn	Sea Goat	*Makara*
11.	Aquarius	Water Bearer	*Kumbha*
12.	Pisces	Fishes	*Meena*

Box 7.1 *Precession of the equinoxes*

We have earlier mentioned in Chapter 6 that the axis of the earth's rotation around itself does not point in a fixed direction. It rotates around a certain direction like the axis of a top, and makes a full circle once in 25 800 years. The perspective is shown in Figure 7.2, which shows the celestial sphere with the ecliptic, the celestial equator, the celestial north pole and the circle of precession of the earth's axis.

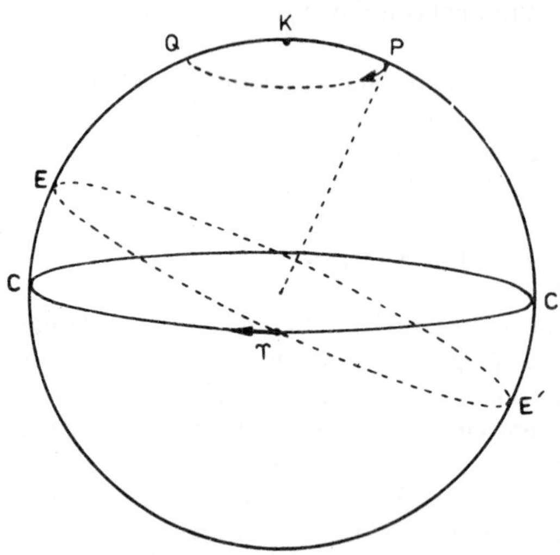

Figure 7.2 The precession of the earth's axis and of the vernal equinox. (Note the slightly different perspective here as compared to Figure 7.1.)

As a consequence, the equinoxes rotate on the celestial equator in the *westward* direction (see also Figure 6.4). This motion is called the *precession of the equinoxes.* Noting that the equinoxes would precess through 360° (24 hr on the time scale) in a little less than 25 800 years, you can see that the precession comes to 50″.29 (3^s.35) per year and to 1° 23′ 49″ (5^m 35^s) per century.

The vernal equinox is the reference point for the astronomical coordinate of right ascension, which is measured eastward. Since the equinoxes precess in the westward direction, the RA of each celestial object increases by 3^s.35 every year, in addition to whatever changes take place due to the proper motion of the object. In fact, it was through this effect that the precession of the equinoxes was discovered and measured about 2000 years ago by Hipparchus.

It is therefore necessary to give the reference date for the equinox while mentioning the coordinates of a celestial object. At present this reference date is chosen to be January 1, AD 2000. So when we say that 'the R.A. of a star (on the equator) is 3^h 12^m 17^s (equinox of 2000)', we should realize that the RA of that star will be less by 33^s.5 on January 1, 1990.

The complete motion is further complicated by the fact that precession of the earth's axis is not the only effect. Like a top, the earth also wobbles around the circle of its precession, a motion which is known as *nutation*. There are also other smaller effects.

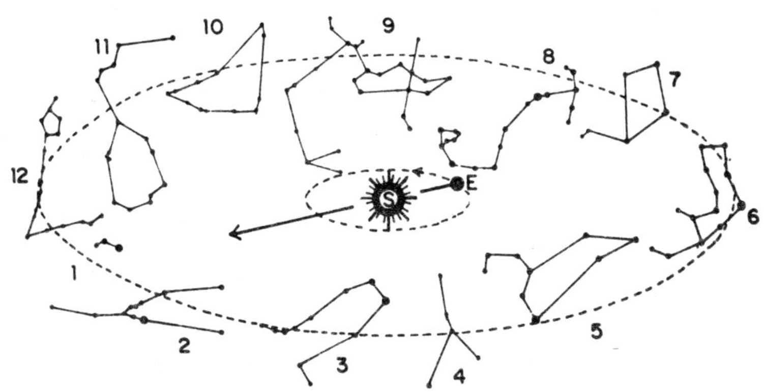

Figure 7.3 The twelve zodiacal constellations in relation to the earth E and the sun S. The numbers correspond to constellations given in Table 7.1.

view. You must think of the zodiac as a circle, with the earth at the centre of this circle, and the twelve constellations arranged serially on this circle. Thus after Pisces, comes Aries.

There are two important results which will come out of this observation, which are really not surprising. These are:

(a) The six visible constellations in the sky at any moment are some six consecutive constellations listed in Table 7.1;

(b) The order of the constellations is from the west to the east.

For example, if you observe Cancer at the western horizon, you will find Leo about 30° above and Virgo about 60° above the western horizon, Libra

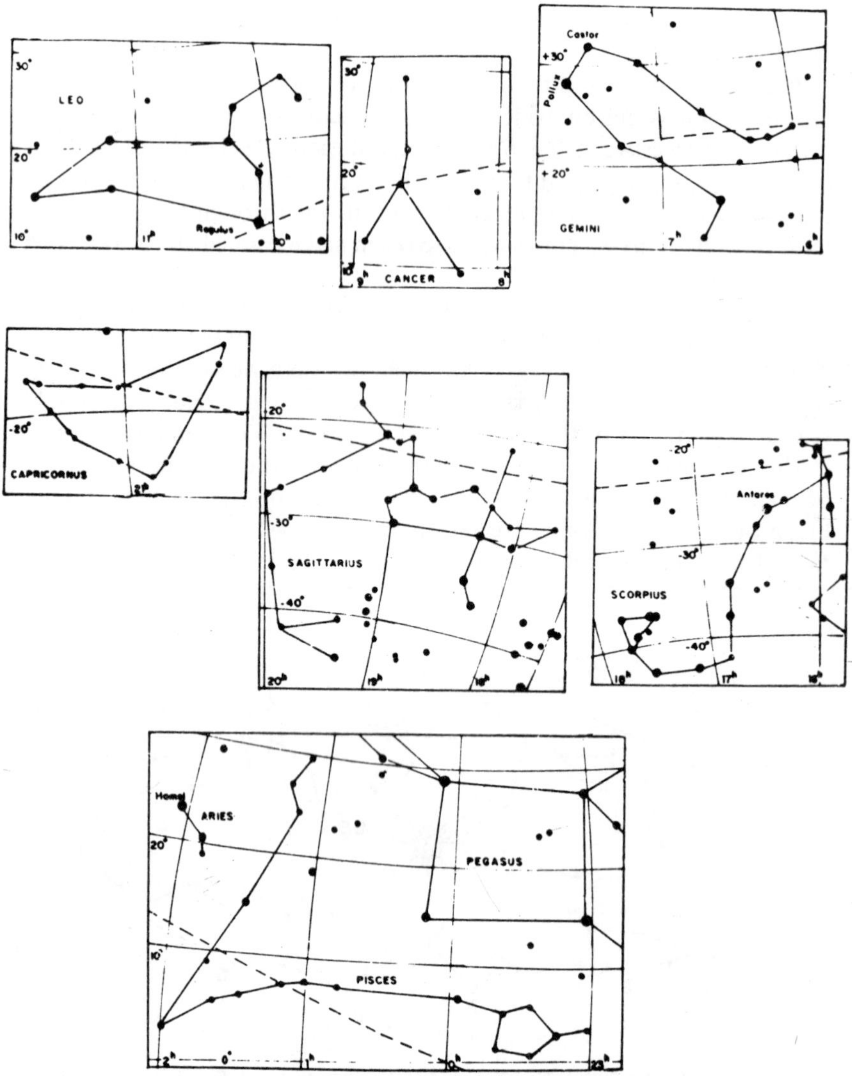

Figure 7.4 The star groups in each of the twelve zodiacal constellations in more detail. The astronomical coordinates, RA and Dec, are shown. The dotted line is the ecliptic along which the sun appears to move. One should hold this figure above one's head with its top pointing to the north. The RA then increases to the left of the page (towards the east in the sky).

near the zenith, and so on until you can find Sagittarius near the eastern horizon. If you watch for an hour or two, you will see Capricorn rising in the east. During the same period, Cancer would have set in the west.

A new constellation rises in the east roughly every two hours and its diametrically opposite star-group sets in the west during the same time. As you should have realised by now, this is the apparent diurnal motion of stars due to the earth's rotation about itself.

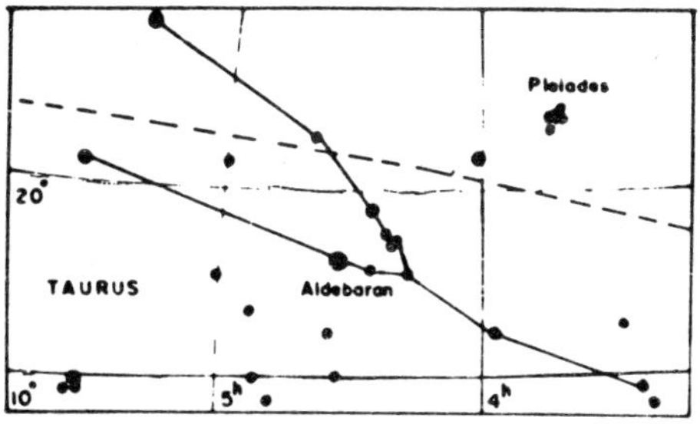

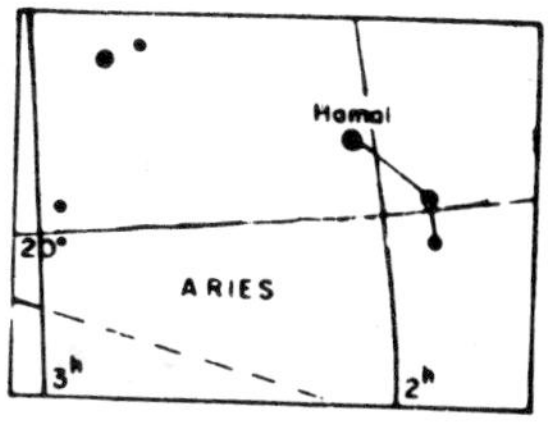

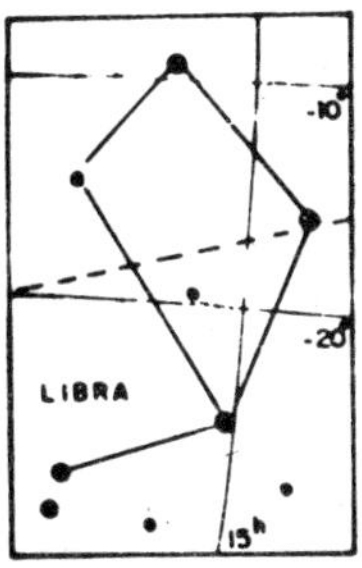

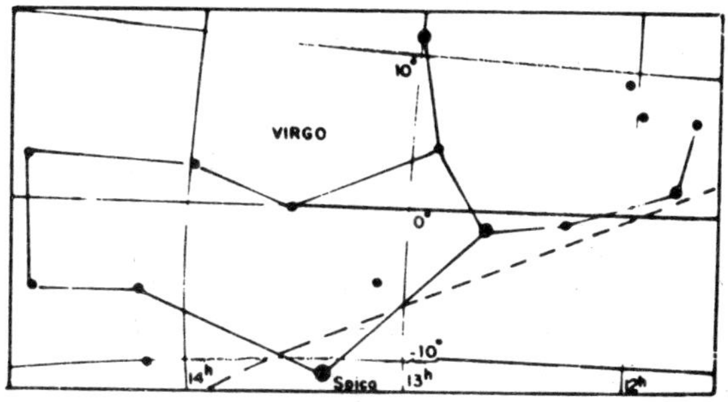

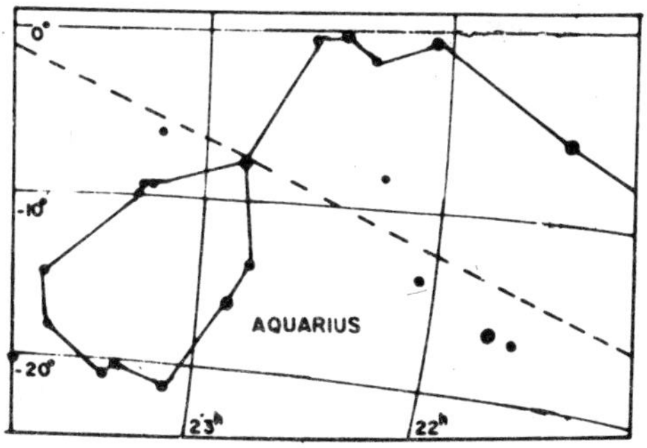

Now you will be able to identify which constellation the moon is in or the constellations the planets are in. You would be able to understand and make statements like, 'Today the moon is in Cancer while Mars is in Virgo'.

Colour Plate 3 is a beautiful photograph showing the conjunction of several planets. This is an event when two or more planets are in the same part of the sky, or in the same constellation.

MOTION OF THE SUN AND THE MOON

Since the moon covers one full circle in its orbit around the earth in about 27 days with respect to the stars, it travels through one zodiacal sign roughly in two-and-a-quarter days.

Box 7.2 *Which constellation is the sun in?*

Can we identify the constellation the sun is in? We cannot do this directly, because stars are not visible during the day. We must look for the constellation near the zenith at midnight. Then the sun will be in the constellation which is diametrically opposite this one. Thus, suppose it is Aquarius (the 11th constellation in Table 7.1) at the zenith at midnight on a particular day. Then the sun must be in Leo (the 5th constellation) on that day.

If you do not wish to wait for the midnight, watch the zodiacal sign at the western horizon two (or four) hours after the sunset. The sun must be in a constellation one (or two) behind the constellation which you see at the western horizon. For example, if the constellation just setting at the western horizon two hours after the sunset is Scorpio (the 8th constellation), the sun must be in Libra (the 7th constellation).

Noting that the moon makes one full orbit around the earth in about 27 days, the ancient Indian astronomers divided the zodiac into 27 parts. Each of them contains a few (sometimes just one) prominent stars. Each of these parts is known as a *Nakshatra*. Some of them are shown in Figure 7.4. These are Hamal (*Ashwinī*), Aldebaran (*Rohiṇī*), Pleiades (*Krittikā*), Castor and Pollux (together *Punarvasu*), Regulus (*Maghā*), Spica (*Chitrā*), Antares (*Jyeshthā*), and Pegasus (*Pūrvā Bhādrapadā* and *Uttarā Bhādrapadā* together).

The daily motion of the moon is better seen in the *Nakshatra* because the moon traverses approximately one *Nakshatra* every day.

Similarly, if you watch the motion of the sun over a few days or months, you will notice that the sun also has an eastward motion relative to the fixed stars of the zodiac. You will further see that it crosses about one constellation per month, and thus covers the entire circle of the zodiac in one year.

The sun crosses over from one constellation to the next roughly between the 14th and the 16th day of every calendar month of the Gregorian calendar (at the present epoch). It is in Aries from mid-April to mid-May, it is in Taurus from mid-May to mid-June, and so on. The transition of the sun from one constellation to the next is known in Sanskrit as *Sankramaṇa.* Every year on January 14, it goes over from Sagittarius (*Dhanu*) to Capricorn (*Makara*). The festival of *Makara Sankrānta* therefore falls on this day every year.

Here again you should realise that this is the apparent motion of the sun which occurs because the earth revolves around the sun once a year. Figure 7.5 shows how this happens.

In this chapter, we have talked a good deal about astronomy. This should be enough for you to get interested in star-gazing. Now we shall take a look at the physical conditions of the objects in the solar system. Let us start with our sun.

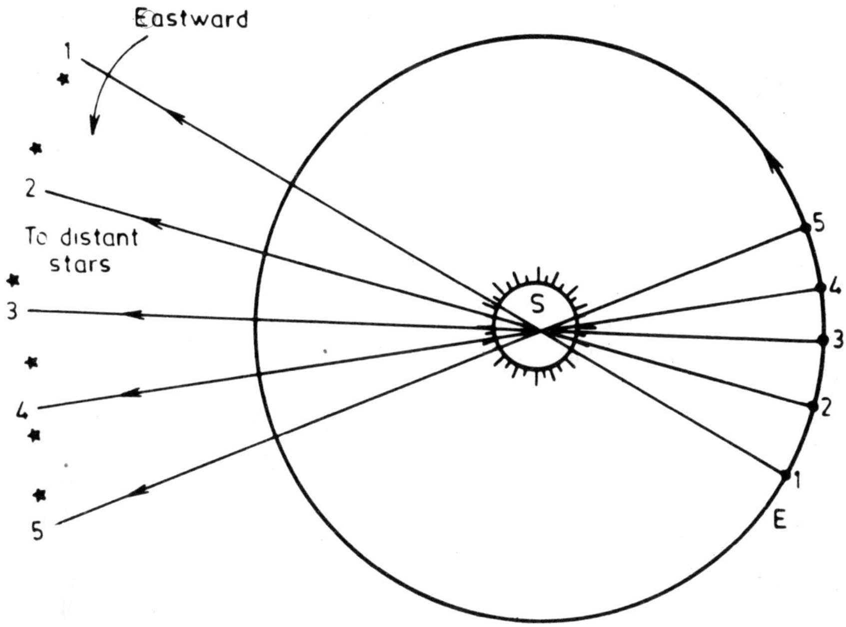

Figure 7.5 Apparent motion of the sun S among the stars. When the earth E is at positions 1, 2, 3, ...in its orbit, the sun S *appears* to be at the respective positions 1, 2, 3, ... in the sky with respect to the stars.

8
The Sun

The sun is the most important object in the solar system. It is the cause of our existence. It provides us most of the energy which we need and consume on the earth.

The sun is a globe of gas. There is no material in the sun in the solid or liquid form, so high is its temperature. It is a ball with a diameter of 1.392 million km (radius 696 000 km). Its angular diameter is 32′ (32 arcmin; see Box 4.1).

In its central parts, the temperature is about 15 million kelvin (see Box 8.1) while it is 6000 kelvin near its surface.

Box 8.1 *The kelvin scale of temperature*

For measuring temperatures in physics, we prefer to use the scale known as *kelvin* and denoted by the symbol K. The temperature in kelvin is obtained by adding 273 to the temperature in degrees celsius. For example, if the temperature of boiling water is 100 °C, it will be (100 + 273) kelvin = 373 K.

The minimum temperature that can be attained in principle is taken as 0 K and is called the *absolute zero*. This temperature would be −273°C on the celsius scale.

The number 273 representing the difference between the two scales is approximate. A more accurate figure is 273.15. This accuracy is not necessary for the purpose of this book.

In Figure 8.1, we have shown a picture of the sun when its disc is totally covered by the moon during a total solar eclipse. You can see the thin radiation coming from the sun which you cannot see otherwise. This is called the *solar corona*.

The density of the central part of the sun is 150 times that of water. This means that one cubic centimeter of this matter would weigh 150 grams (on the earth) whereas one cubic centimeter of water on the earth weighs one gram. This happens due to the high gravitational compression of the sun leading to very high pressure, density and temperature in the core.

The mass of the sun is estimated to be 1.9891×10^{30} kg, or approximately 2×10^{30} kg. (Place 30 zeros after 2 and you will get this number.) This

Figure 8.1 Solar corona visible during a total solar eclipse. (*Courtesy*: UPSO)

is about 332 946 times that of the earth. This can be found from the time of revolution of the earth around the sun and the distance between the two (see Box 5.1).

The visible surface of the sun is called the *photosphere*. It is this photosphere from which light and other high energy particles are released by the sun in all directions. A thin shell of thickness about 2200 km between the photosphere and corona is called the *chromosphere*. It is given this name because many strong colourful emissions take place at special wavelengths from this region. (*Chrome* in Latin means colour.) The chromosphere can be seen with the naked eye only during a total solar eclipse.

The sun also rotates around itself once in about 25 days. But it is not a solid body—it is gaseous. Therefore its different parts can rotate with different speeds. There are times when the sunspots are seen with the naked eye shortly before the sunset (see Box 8.2). By noting the shift of their positions, you can easily convince yourself that the sun is in fact rotating about its own axis.

The axis of rotation of the sun is inclined to the plane of the ecliptic by about 83°, or it is inclined to the normal to the plane of the ecliptic by about 7°. The perspective is shown in Figure 8.2.

The sun provides us with most of the energy we need for survival. In fact, life has evolved on the earth because of this constant supply of energy from the sun.

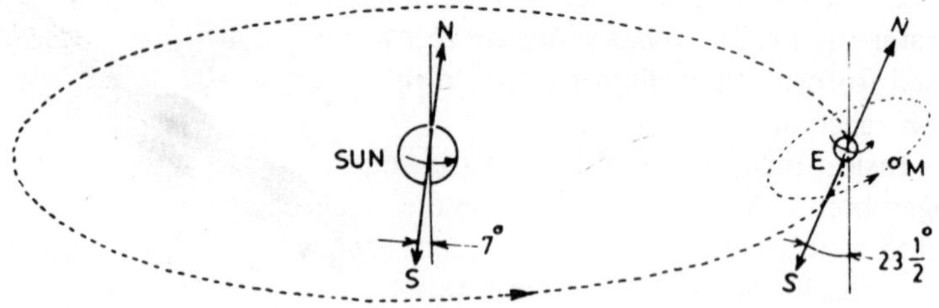

Figure 8.2 The axis of rotation of the sun around itself makes an angle of 7° with the normal to the plane of the ecliptic. The earth's orbit around the sun and the moon's orbit around the earth are also shown.

CONTENTS

The sun contains mainly hydrogen, some helium and a very small fraction of other chemical elements. The contents of various elements in the sun are shown in Table 8.1. You will see from this table that hydrogen and helium make for 98% of the sun and all the other elements come to only 2% by mass.

ENERGY PRODUCTION

How does the sun or any other star produce such vast amounts of energy? The answer to this question has been found only after 1930.

It is now believed that a star contains mainly hydrogen, some helium and a small fraction of other chemical elements. The composition of an average star is like that of the sun, which is given in Table 8.1. Note that hydrogen is the lightest element while helium is the next lightest element in the world. All other elements are heavier that these two.

Table 8.1 The contents of the sun

Element	*Percentage by mass*
Hydrogen (H)	70.52
Helium (He)	27.57
Oxygen (O)	0.96
Carbon (C)	0.31
Iron (Fe)	0.18
Nitrogen (N)	0.11
Silicon (Si)	0.07
Magnesium (Mg)	0.06
All other elements together	Less than 0.22

The large mass of the sun results in a high pressure at the centre due to the heavy weight of the overlying material, which can be balanced only if the temperature in its central parts becomes high. A new type of reaction takes place at these temperatures. In this reaction, hydrogen nuclei get converted into helium nuclei. Since such a nuclear reaction takes place at a high temperature, it is called a *thermonuclear reaction.* In effect, four hydrogen nuclei get fused to form one helium nucleus in this reaction—hence it is also called a *fusion* reaction.

This fusion reaction releases a vast amount of energy. It is as if several hydrogen bombs were exploding inside the sun every second, converting 4.25 million tons of hydrogen into helium each second. So this is how energy is being produced in the sun and in any star.

You will see from the above that the stock of hydrogen in the sun is being consumed every moment and helium (and possibly other elements) are being created. You can then imagine that some time or the other hydrogen in the sun will get exhausted. There will be no more fusion and the sun will stop supplying the energy it now does. This will be the death of the sun, and also of the earth.

However, you need not worry about this happening in the near future. It is believed that the sun has existed for about five billion years and will continue to exist for another eight or nine billion years.

SUNSPOTS AND MAGNETIC FIELD

Observation of the sun even with a telescope of moderate size shows several dark spots on its surface. The number of spots and their size vary from year to year. Figure 8.3 shows the sun revealing a few sunspots. Figure 8.4 is a high-resolution photograph of a large group of sunspots.

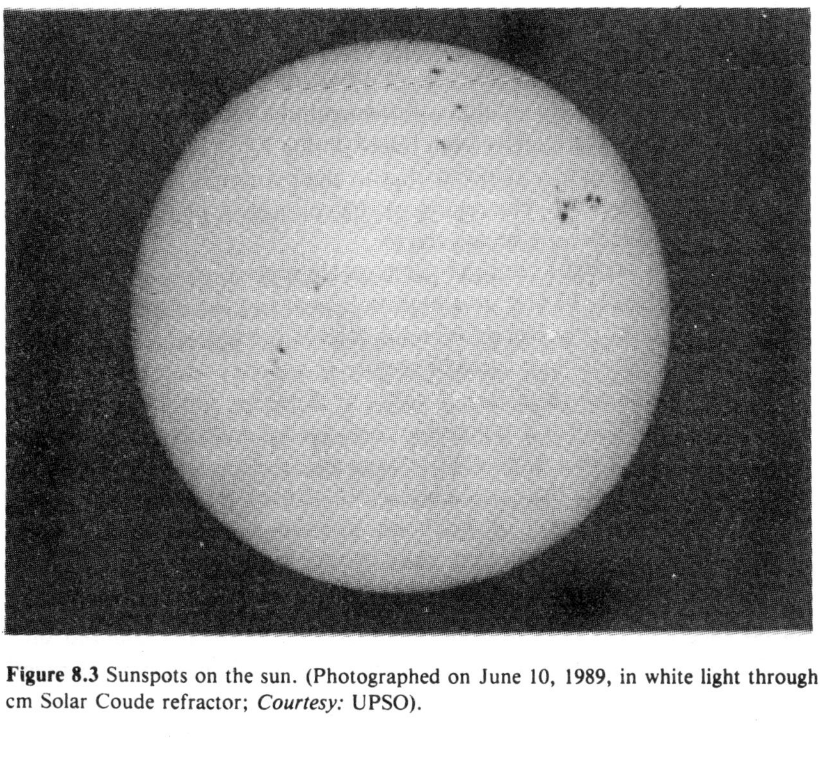

Figure 8.3 Sunspots on the sun. (Photographed on June 10, 1989, in white light through 15 cm Solar Coude refractor; *Courtesy:* UPSO).

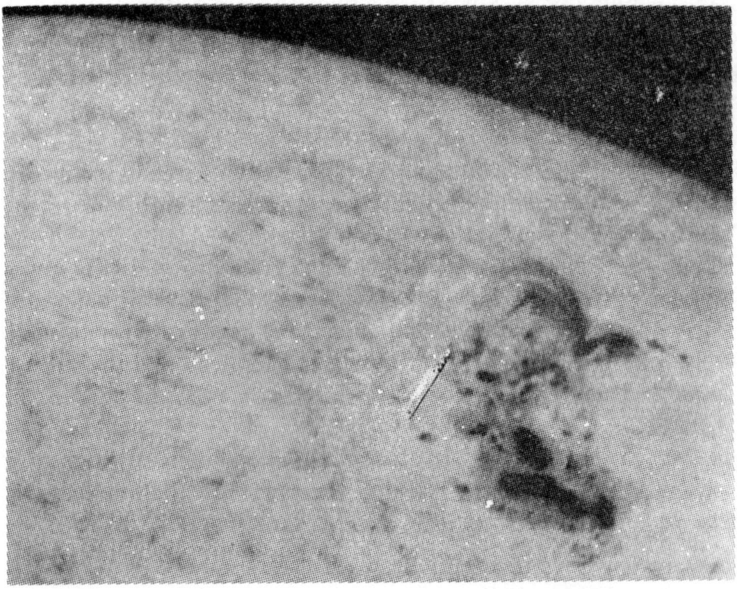

Figure 8.4 A high resolution 'close-up' of a group of sunspots. The linear dimensions of the dark group of sunspots is about 70 000 km left-to-right and over 80 000 top-to-bottom. (Photographed on June 13, 1982, at a wavelength of 6563 angstroms of H_α radiation; *Courtesy:* USO)

Close observations over a number of years have shown that the growth of sunspots occurs in a periodic cycle of 11 years. Thus at the end of every period, the sunspot activity would be at a maximum. During the recent years, peaks in the sunspot activity have been found during 1957, 1968, 1979 and 1990.

It is believed that sunspots occur due to the extremely intense magnetic field produced in the sun. The region of the sun near a high magnetic field gets cooler and therefore appears dark.

The sun has electrically charged particles such as electrons, protons and ions. In a magnetic field and at a high temperature, these particles would move along complex trajectories forming interesting patterns. When this is coupled with the eleven-year cycle of magnetic activity, you can see that the solar corona will produce different views at different times.

Compare the solar corona appearing in Figure 3.1 with that in Figure 8.1. In the former case, it is a symmetric corona and indicates low magnetic activity. In the latter case, the corona consists of *streamers* and indicates high magnetic activity. The cover of this book also shows a solar corona of the latter type. It is a picture of the sun taken by the Skylab space station during a total eclipse. The streamers appear due to the sun's magnetic field and extend far into the interplanetary space.

The nature of sunspots and magnetic field of the sun are not yet very well understood. A good deal of research is going on in this area in different parts of the world. New and surprising results prop up every now and then.

PROMINENCES AND FLARES

If you think that the sun's surface must be a place of great dynamic activity, you are right. Radiation, which consists of light and several particles, is produced throughout the interior of the sun. It travels from the interior to the surface. It is from the surface (photosphere) that all this radiation is spewed out into the space.

Therefore the surface of the sun is not smooth like that of a ball. Figure 8.5 shows a *prominence* on the surface of the sun. This is a large eruption of gas from some part. An abrupt increase in the emission of light from a

Box 8.2 *Observation of the sun*

Clearly the sun cannot be observed with the naked eyes due to its brilliance. Different techniques have been developed by physicists to learn about the various aspects of the sun.

The first and simple instrument is a telescope. You can find a small telescope in the science department of almost any college or university. Science museums and planetaria also have such telescopes. You can observe some features of the sun with these telescopes. However, you should take care *not to look at the sun through a telescope* even if you get hold of a dark solar filter lest you should damage your eyes. Focus your telescope on a distant tree before you turn it towards the sun. Hold a piece of paper about half a metre away from the eyepiece. You will see the image of the sun as an illuminated disc with sunspots on it.

For more accurate observations, however, we require large telescopes. The larger the lens or mirror of a telescope, the smaller the details you will be able to see through it. A telescope is therefore known by its size—that is, the diameter of the lens or mirror that directly faces the object, called the *objective lens* or simply the *objective*.

Apart from this, the sun can be observed in infrared and ultraviolet radiation, as well as radio waves. There are special instruments each of which is sensitive to a particular kind of radiation. See Chapter 18 for more detail.

It is also possible to study various kinds of particles coming from the sun. By observing particles like electrons, protons, neutrinos and other ions coming from the sun, it has been possible during the past few years to know something about the interior of the sun.

Observations of the sun during a total solar eclipse have their own importance. They yield a good deal of information on the solar emissions from its atmosphere. That is why scientists from all parts of the world flock, along with their instruments, to a suitable place from where the total eclipse can be observed surely and easily.

Several manned and unmanned spacecraft have also been used to study the sun. Observations from a spacecraft have the advantage that radiation falling on it does not have to pass through the earth's atmosphere.

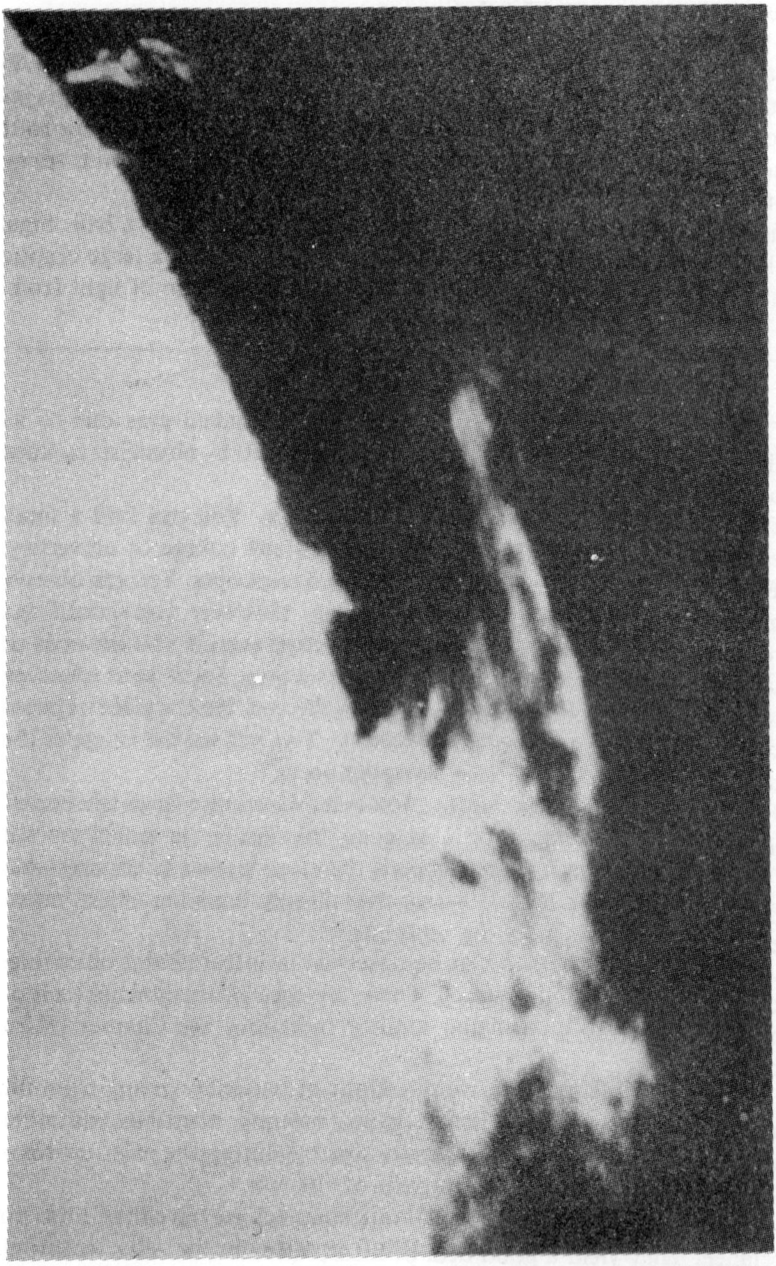

Figure 8.5 A high resolution photograph of a prominence taken on May 30, 1989, in the H_α line. The top edge of the prominence is at a height of 72 000 km from the solar limp. (*Courtesy: USO*)

sunspot region is called a *flare*. Figure 8.6 shows a two-ribbon flare, so known due to its appearance, in high resolution. The two bright curvilinear ribbons are separated by dark filaments. One can also see bright post-flare loops on the top side. The flare occurred in a complex sunspot group near the east limb of the sun.

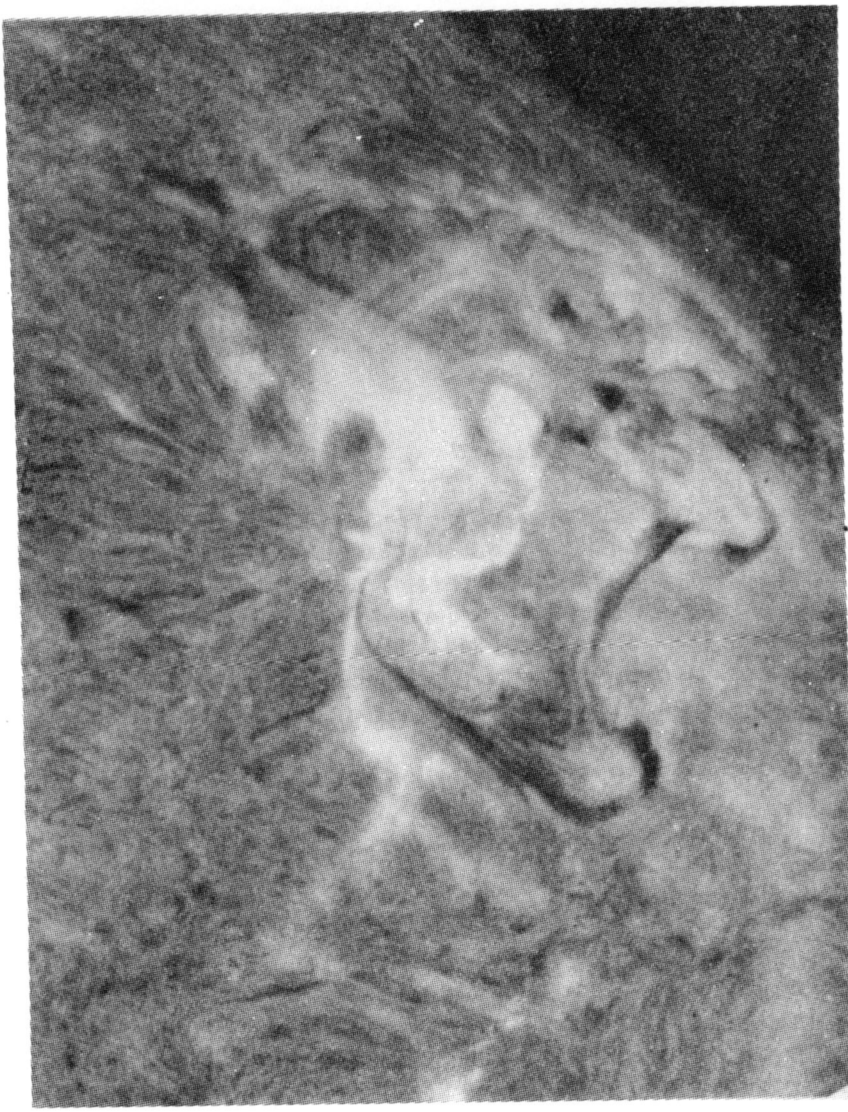

Figure 8.6 A high resolution picture of a great two-ribbon flare observed in H_α line on May 13, 1981. The linear extension (top-to-bottom) of the flare is about 200 000 km. (*Courtesy*: USO)

Figure 8.7 is a photograph taken by eclipsing the sun's disc artificially. It clearly shows solar prominences and flares. Prominences and flares can sometimes be very colourful.

Figure 8.7 Flares and prominences on the surface of the sun photographed by eclipsing the sun by a disc. (*Courtesy*: IIA)

HISTORICAL DEVELOPMENT

Until about five centuries ago, people thought that the earth was at the centre of the universe, and that it was stationary. Everything else including all the stars and the sun were supposed to move around the earth. (This is known as the *geocentric system*.) This view was changed due to careful studies by Copernicus, Tycho Brahe, Kepler and Galileo. Since then it is generally agreed that it is planets of the solar system, including the earth, which go round the sun. In this model, the sun is relatively stationary. This is known as the *heliocentric system*.

It is much easier to explain the motion of the planets in the heliocentric system. In the geocentric system, their motion would appear to be haphazard. (See discussion about forward and retrograde motion of a planet in Chapter 4.) It would be difficult to explain this motion on the basis of any systematic scientific law.

SUN'S MOTION IN THE GALAXY

The sun, along with its entourage of planets, satellites, asteroids, comets and what not, is believed to move in our Galaxy with a tremendous speed. Recent careful observations of the motion of nearby stars have enabled astronomers to estimate the motion of the sun. It is now believed that the solar system is moving with a speed of about 220 km per second towards the direction of the Cygnus constellation.

We shall say more about this when we talk of the place of the solar system in the universe. (See Chapter 19.)

9

The Moon

The earth has only one natural satellite, the moon. It is the most beautiful object to watch in the night sky. Like the earth, it rotates about itself. It also revolves around the earth.

The moon is almost spherical in shape, with a radius of about 1738 km (diameter 3476 km). Its angular diameter is 31′, which is nearly the same as that of the sun. The mass of the moon is 7.3483×10^{22} kg, which is about 1/81.3 times that of the earth.

Colour Plate 4 shows the full moon. Figure 9.1 is also a photograph of the full moon taken from Apollo 17.

Figure 9.1 The full moon. (Photograph taken from Apollo 17; *Courtesy*: TG)

Box 9.1 *Apparent sizes of the moon, the sun and a coin*

It is seen that the apparent size of the moon is almost the same as that of the sun. This is evident from the fact that the moon can just cover the entire disc of the sun during a total solar eclipse.

Let a_M and a_S be the distances of the moon and the sun respectively from the earth, and d_M and d_S be their actual diameters, respectively. The fact that their apparent angular diameters are the same tells us that

$$\frac{d_M}{a_M} = \frac{d_S}{a_S}.$$

Knowing three of the above quantities, we can calculate the fourth. (See also Box 4.1.)

How does the apparent size of the moon or the sun compare with that of a one rupee coin? Ask your friend to hold a one rupee coin between the moon and your eyes on a full moon night. Ask him to hold it such that its plane is perpendicular to your line of sight of the moon. Closing one eye and moving forward or backward, see at what distance the coin must be kept from your eye so that it may just eclipse the entire moon. For a coin of diameter 2.8 cm, you will see that this distance comes out to be 3 m. Thus the apparent size of the moon or the sun is the same as that of a one rupee coin at a distance of 3 m. See Figure 9.2.

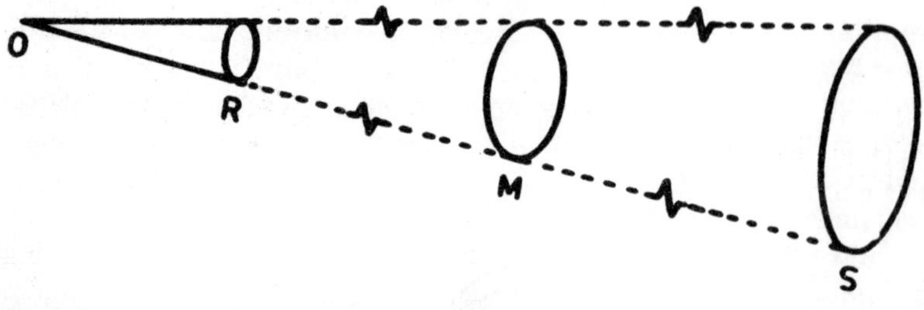

Figure 9.2 The extremities of the moon (M) and the sun (S) subtend the same angle at the eyes of an observer (O) on the earth as a one rupee coin (R) at a distance of 3 m.

The surface gravity on the moon is one-sixth the gravity on the earth. This means you will be able to make a high-jump or long-jump on the moon six times as high or as long as on the earth (assuming that you can produce the same amount of thrust on the moon as on the earth). If you use a spring balance to weigh objects, the weight of every object on the moon will be only one-sixth of its weight on the earth.

The moon revolves around the earth in 27 days 7 hours 43 minutes 11.6 seconds (about 27.3 days). It thus covers about 13°.2 of the sky every day (180° in 27.3 days). Its orbit around the earth is fairly elliptical. This orbit is inclined to the plane of the ecliptic at a small angle of a little over 5°. That is, the perpendicular to the moon's orbit makes an angle of about 5° with the perpendicular to the plane of the ecliptic. The perspective is explained in Figure 9.3. Note that the earth's motion is shown *with respect to the sun* while the moon's orbit *with respect to the earth.*

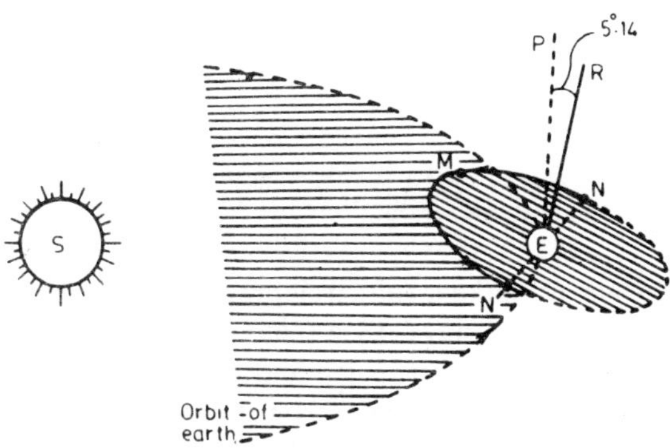

Figure 9.3 The orbit of the moon (M) around the earth (E) is inclined to the plane of the ecliptic by about 5°. EP is the normal to the plane of the ecliptic and ER is the normal to the moon's orbit. N and N′ are the nodes. S is the sun. ER itself rotates about EP.

The average orbital speed of the moon in its path around the earth is 1.02 km/s *with respect to the earth.*

Observations have shown that the moon always shows us the same face. This indicates that it rotates around itself in about the same time as it takes to make a revolution around the earth.

You can understand this from an analogy. It is like a person going around a deity in a temple, keeping the deity always to his right (that is, offering a *pradakshina*). As he goes around the deity once, he also completes a rotation about himself. Similarly, the period of rotation of the moon about itself and the period of its revolution around the earth are exactly the same. Hence we can see only one side of the moon. See Figure 9.4.

In reality, the speed of revolution is not uniform due to the ellipticity of the orbit, though the speed of rotation is perfectly uniform. Hence we can peep through to some parts of its rear side. In fact, the moon shows us about 59% of its surface.

There is another interesting consequence of the near-equality of the periods of rotation and revolution of the moon. From a particular place on the moon, the earth will always appear at the same angle in the sky above the moon's

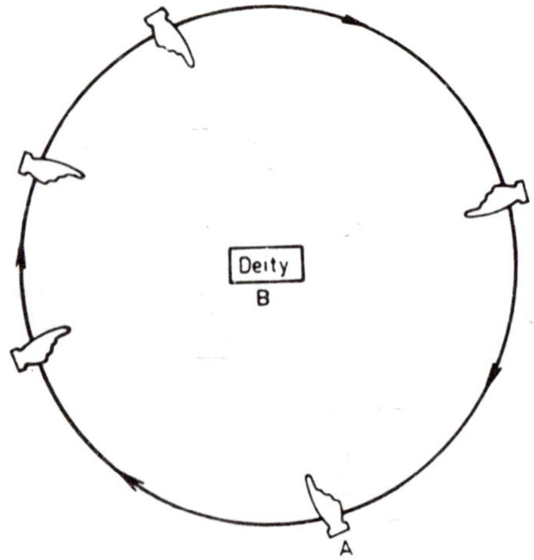

Figure 9.4 Body A turning around itself as well as going around Body B in the *same* duration of time. An observer on Body B can see only the same half the surface of Body A.

horizon. It will appear to make a full circle, remaining at the same altitude in the sky, once in about 27 days. (See Colour Plate 1.)

The average distance between the earth and the moon is 384 400 km. But the orbit of the moon around the earth is not exactly circular. Hence the earth-moon distance varies by about 50 000 km. When the moon comes nearest to the earth, it is at a distance of about 356 000 km, while at its farthest, it is at a distance of about 406 000 km (which is a 13% variation). Even these minimum and maximum distances of the moon from the earth vary slightly during each revolution.

Since the length of a 'day' on the moon is about 13.66 earth-days, its day-side becomes very hot and the night-side rather cold. The highest temperature on the surface of the moon at its equator during a lunar day reaches as high as 130°C while the night temperature on the equator shoots down to about −180°C. The temperature at its poles is about −130°C.

It has been found from careful observations that the revolution of the moon around the earth is *slowing down.* This means that the period of revolution is increasing. At the same time, the average distance of the moon from the earth is also increasing. This effect occurs because of the friction produced by tides which are themselves caused by the moon. The rate of increase in the period of revolution is extremely small, about 0.014 second per century.

PHASES OF THE MOON

We have already discussed in Chapter 4 how the moon shows various phases. Here we shall mention some other aspects of the lunar motion.

The period between two successive new moons is called a *synodic* lunar month. It is this motion on which all traditional Indian calendars are based. A new moon occurs when the moon comes between the earth and the sun. The lunar month begins immediately after the new moon. Then the moon goes through a period of increasing phases till the full moon day, followed by a period of decreasing phases, and finally the new moon day again. This whole period is one synodic lunar month.

It has been found that one synodic lunar month consists of 29 days, 12 hours, 44 minutes, and 3 seconds. Remember that the period which the moon takes for making one revolution of the earth is 27 days, 7 hours, 43 minutes, and 12 seconds, which is called the *sidereal* lunar month. Why does this difference between the two periods occur?

This is because the period of revolution around the earth is measured with respect to the stars. The moon comes back to the direction of the same star after $27^d\ 7^h\ 43^m\ 12^s$. On the other hand, the period of one synodic lunar month is measured with respect to the sun, as the moon has to return to the direction of the sun from new moon to new moon. These two periods are different because the earth is itself moving around the sun, and it carries the moon with it during this motion.

In fact, we have already explained this aspect in Chapter 6 (Box 6.1) in connection with the combined rotation and revolution of the earth. The same thing happens in the case of the moon. Consider that the earth is stationary with respect to the sun and that only the moon revolves around the earth. In this situation, the period of revolution of the moon with respect to the sun or the stars would be the same, that is $27^d\ 7^h\ 43^m\ 12^s$. But during this period, the earth goes around the sun. Hence the perspective changes with respect to the sun and the moon takes $29^d\ 12^h\ 44^m\ 3^s$ to come back to the direction of the sun. Note that the sun moves by $29^\circ.105\ 62$ in $29^d.530\ 589$ and the moon takes $2^d\ 5^h\ 0^m\ 51^s$ over its sidereal period in order to complete one synodic period.

This is explained in Figure 9.5. In this figure, A, B and C are the positions of the earth E in its orbit around the sun S, and M is the moon. The period

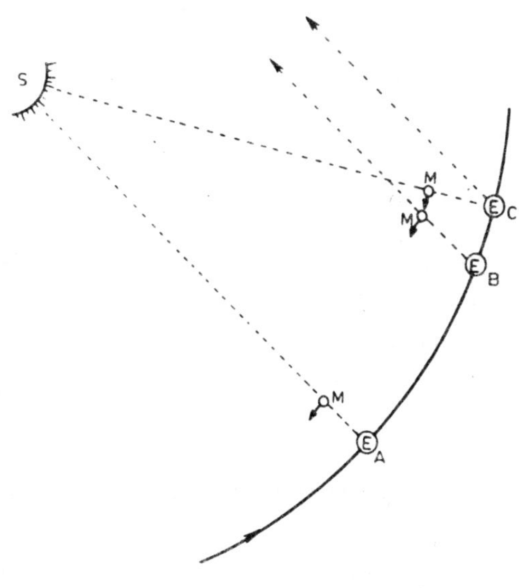

Figure 9.5 Difference between the period of revolution of the moon (sidereal month) and the synodic lunar month. (Arrows show the motion of the earth *with respect to the sun,* and that of the moon *with respect to the earth.*)

between the positions A and B is the sidereal lunar month, while that between A and C is the synodic lunar month.

OBSERVATIONS OF THE MOON

The moon is the only celestial body other than the earth on which man has been able to set foot. Many unmanned spacecraft have also been sent towards the moon. Some of these have been put into orbit around the moon and some others have landed on it. Naturally, we know more about the moon than any other celestial body.

Two astronauts from the earth first landed on the moon on July 20, 1969. The landing site of the first humans on the surface of the moon is shown in Figure 9.6. Several other astronauts have landed on the moon since then. In addition, several astronauts have orbited the moon in their spacecraft at close distances. These astronauts as well as the unmanned spacecraft have brought back samples of the lunar soil, rocks, a large number of photographs and a wealth of other information. These rock samples and photographs were studied in many laboratories all over the world.

STRUCTURE OF THE MOON

It has been conclusively proved from these observations that there is no atmosphere on the moon. It was a long-standing question whether microorganisms exist in the lunar soil. No trace of any organisms was found in the analysis of lunar soil from different regions of the moon. There is also no water or water vapour on the moon.

Box 9.2 *Is there really no atmosphere on the moon?*

When we say there is 'no' atmosphere on the moon, you should understand that it is *not really zero*. It means that there are too few molecules of any gas in a given volume on the surface of the moon as compared to those on the earth.

At sea level on the earth, the average pressure is one atmosphere. Under such conditions and at the normal temperature of 20°C, you would have approximately 2.7×10^{25} molecules of all gases together in a cubic metre. On the surface of the moon, the pressure is about 10^{-11} to 10^{-12} times that on the earth. This means that there would be between 10^{14} and 10^{13} molecules per cubic metre on the surface of the moon at 20°C.

The moon's surface has several mountains as on the earth. There are also several craters. Some of the craters are very large. Figure 9.7 shows the Alps crater whereas Figure 9.8 shows the Tycho and Clavius craters on the moon's surface.

Figure 9.6 The landing site of Apollo 11 astronauts on the moon. (*Courtesy*: TG)

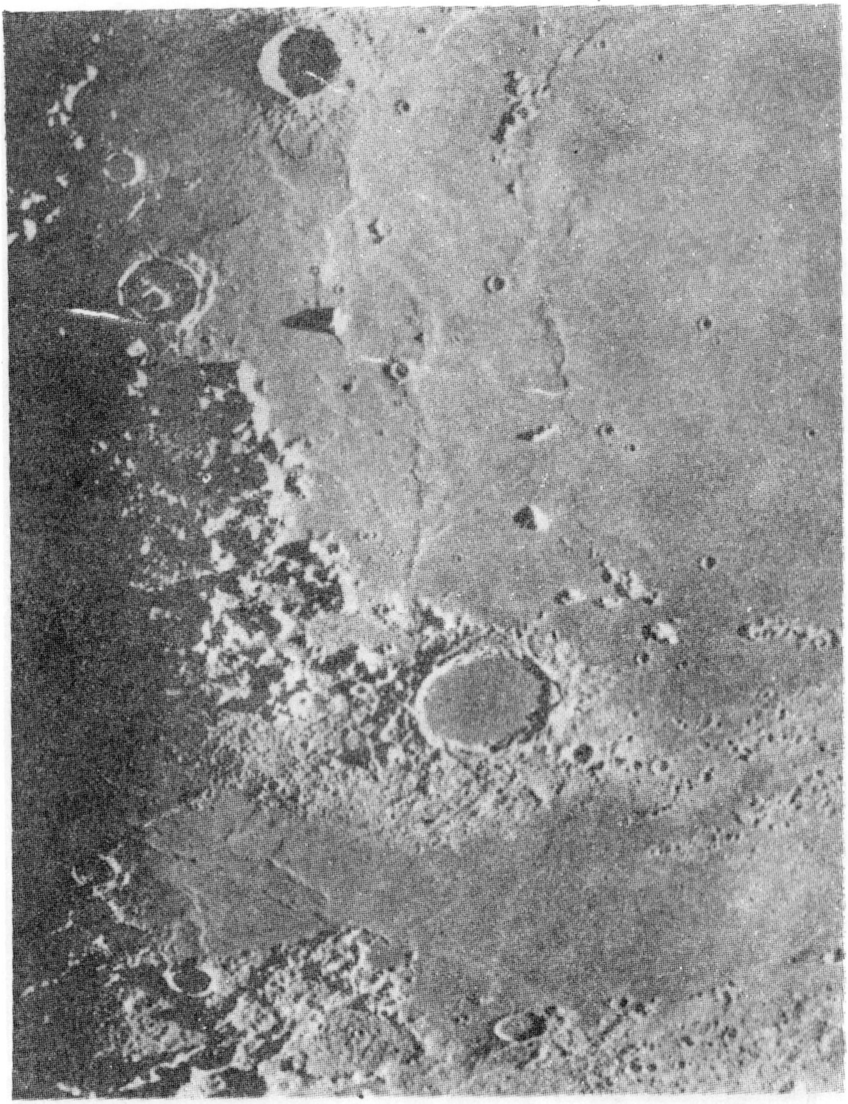

Figure 9.7 Alps crater on the moon's surface. (*Courtesy*: LO)

We have remarked earlier in Chapter 6 that the earth's atmosphere protects us from deadly ultraviolet radiation coming from the sun as well as from the shower of meteoroids flying in space and colliding with the earth and becoming meteors. Since the moon has no atmosphere, its surface is subjected to both these hazards. There is no life on the moon, because of the ultraviolet radiation from the sun and absence of water, if not for anything else. The craters on the moon have been created owing to the impact of large meteoroids on it.

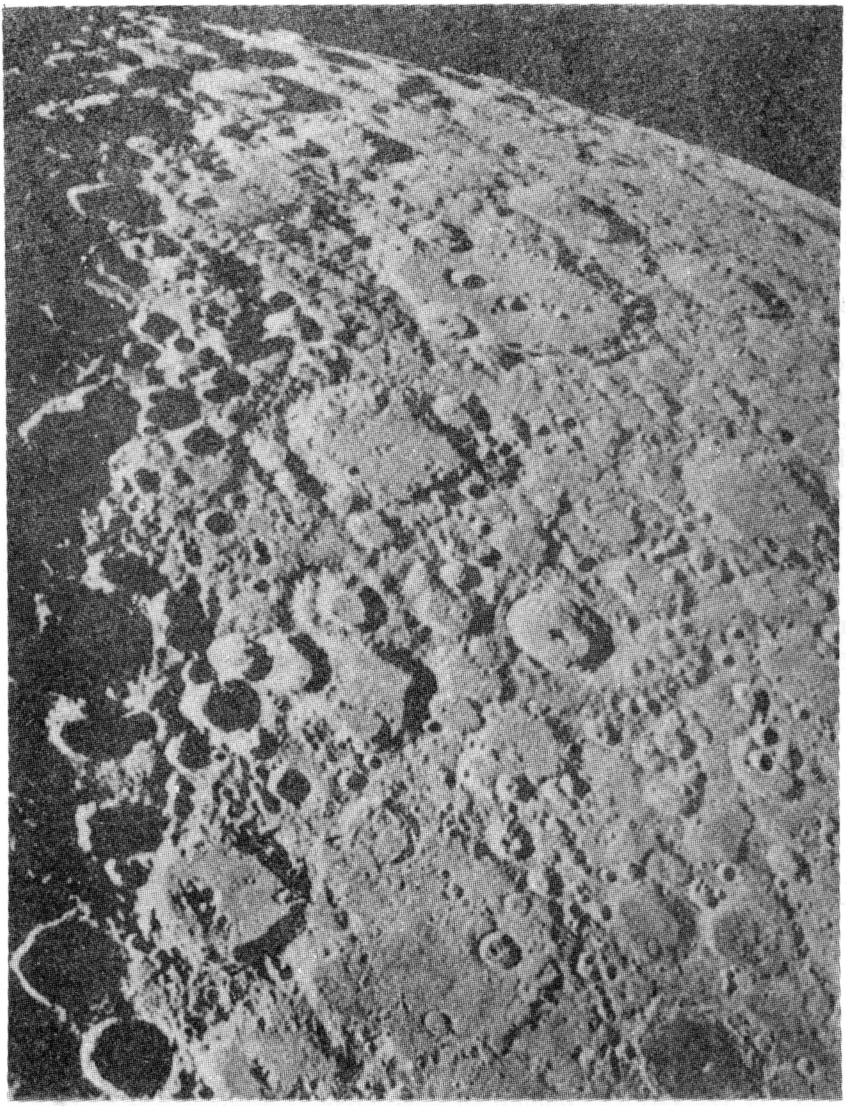

Figure 9.8 Craters Tycho and Clavius on the moon. (*Courtesy*: LO)

The moon's soil and rocks are roughly similar to those on the earth. Astronauts have placed seismometers (see Box 5.2) on the surface of the moon. They have shown that the moon also has moonquakes. However, not much is known about the interior of the moon, except that it is solid throughout.

10

The Sun-Earth-Moon System and Eclipses

The sun and the moon are the two members of the solar system (and of the cosmos at large) which affect our daily lives to a great extent. We shall therefore study the geometrical and astronomical relationships among the three bodies in some detail. In particular, we shall explain how and when eclipses of the sun and the moon occur.

THE EARTH-MOON SYSTEM

We have already mentioned that the earth is about 81.3 times heavier than the moon. The centre of mass (CM) of the earth-moon system therefore lies at a point close to the centre of the earth on the line joining the earth to the moon. As explained in Box 10.1, the CM divides the earth-moon distance in the ratio 1:81.3.

Box 10.1 *Centre of mass of the earth-moon system*

Consider two spheres A and B, of different masses, joined by a thin rod. Now take a string and tie one end at some point P near the centre of the rod. This is shown in Figure 10.1 (a). Hold the other end of the string and lift the rod along with the spheres. The heavier sphere will go down and the lighter one will go up, as shown in Figure 10.1 (b).

Now move the point P, where the string is tied to the rod, toward the heavier sphere. In a few trials, you can find a suitable point C such that the rod remains horizontal even when lifted. This point C is called the *centre of mass* of the system consisting of the two spheres and the thin rod. See Figure 10.1 (c). (Here we have assumed that the gravitational field is uniform over the system consisting of both the spheres and the rod.)

Now imagine that the rod is made thinner and thinner. Every time, find the CM of the system. Finally imagine that the rod vanishes and we have only the two spheres. Even in this situation, we say that C is the centre of mass of the system consisting of the two spheres. This is shown in Figure 10.1 (d).

Suppose we look at the system of Figure 10.1 (c) from the top. Set the two balls in rotatory motion so that the rod connecting them remains horizontal and the path of each ball is circular with C as the centre.

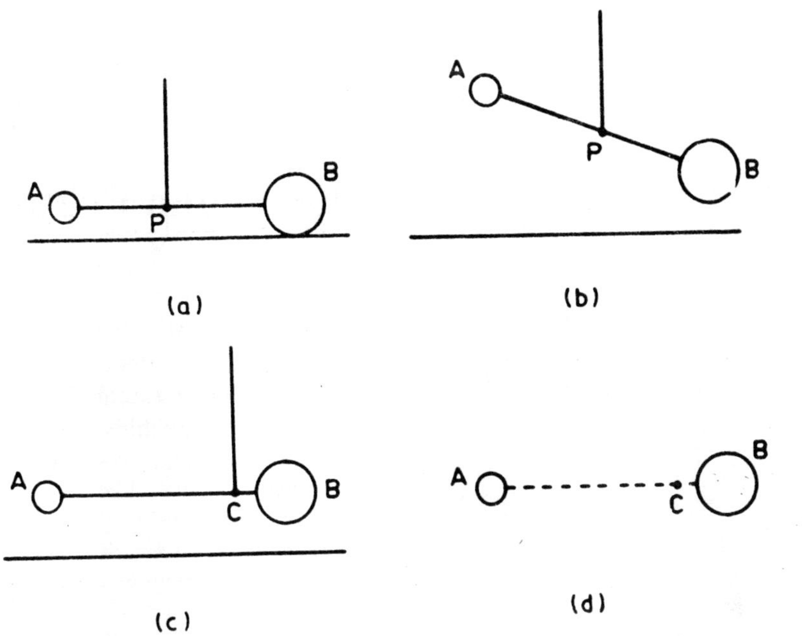

Figure 10.1 Illustrating the concept of the centre of mass of two balls, one light and the other heavy.

This is shown in Figure 10.2. By the time the smaller sphere A goes to the point A′, the larger sphere B will go to the point B . The rod connecting them always remains horizontal and their centre of mass C remains stationary.

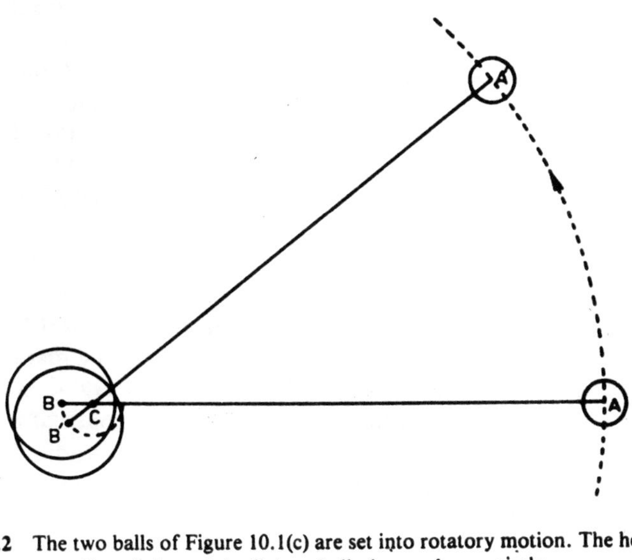

Figure 10.2 The two balls of Figure 10.1(c) are set into rotatory motion. The heavier ball moves along a smaller circle and the lighter ball along a larger circle.

Let the mass of the small sphere A be *m* and that of the large sphere B be *M*. You will see that the centre of mass C lies closer to the larger mass B. In fact, the distance of the CM from the centre of each sphere is inversely proportional to their masses. Thus, BC/AC equals *m/M*.

In the case of the earth-moon system, let A be the moon and B be the earth. Then the ratio BC/AC would be 1/81.3, or the ratio BC/AB would be 1/82.3.

Since the average earth-moon distance is 384 400 km, the distance of their CM from the centre of the earth would be 384 400 km/82.3 which is about 4666 km. Since the radius of the earth is about 6370 km, this point lies inside the earth. The situation is very similar to that shown in Figure 10.2 and explained in Box 10.1, except that the CM of the earth-moon system lies *inside* the earth.

Therefore as the moon revolves around the earth, the earth must also move along a path around the common CM. Instead of saying that 'the moon revolves around the earth', we should correctly say that *the earth and the moon revolve around their common centre of mass,* known as the *barycentre* of the earth-moon system.

MODIFICATION DUE TO THE SUN

This would be the case if there were only the earth and the moon in the whole universe. But this earth-moon system revolves around the sun once a year. In that case, what would the path of the earth around the sun look like?

Instead of following the ideal path shown by the continuous line in Figure 10.3, the centre of the earth would follow the path shown by the dotted line. In fact, the centre of mass of the earth-moon would always lie on the continuous line. This produces a small variation in the sun-earth distance during each month. The earth is closer to the sun on a full moon day than on a new moon day by about 9332 km. Thus it causes wiggles in the motion of the earth around the sun. These wiggles have indeed been observed. This variation is in addition to the variation due to the fact that the earth's orbit is not really circular.

It was through these effects that the mass of the moon was estimated in the early days of astronomy. We now have a much more accurate value for its mass from the observations of artificial satellites which have been put in orbit around the moon (see Box 5.1).

Note that the wiggle of the earth about the elliptic path of the earth-moon CM has been shown in Figure 10.3 in a highly exaggerated manner. In fact, the earth's path with respect to the sun is *always concave* towards the sun. If the sun-earth distance is taken as 15 m (about the length of a hall), then the average earth-moon distance would be 3.84 cm. The radius of the earth would be 0.637 mm and the CM of the earth-moon system would be at a distance of 0.467 mm from the earth's centre. The wobble in the sun-earth distance on this scale would come only to 0.933 mm or less than 0.1 cm.

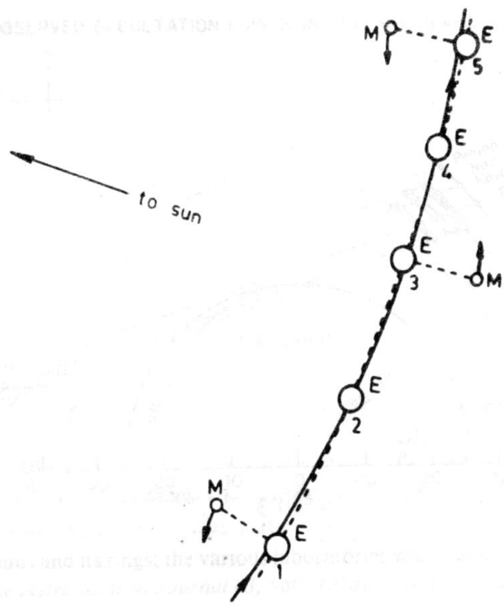

Figure 10.3 The wiggly motion of the earth around the sun due to the presence of the moon. E is the earth and M the moon. (Arrows indicate the direction of revolution of the earth *with respect to the sun* and that of the moon *with respect to the earth;* highly exaggerated and not to scale.)

INCLINATION AND PRECESSION OF THE MOON'S ORBIT

Now we come to the combined system of the three bodies, the sun, the earth and the moon. We have mentioned in the previous chapter that the plane of revolution of the moon around the earth is not the same as the plane of revolution of the earth around the sun (which is known as the plane of the ecliptic). The orbit of the moon is inclined to the plane of the ecliptic by an angle of about 5°. (It really varies over a range from about 4° 59′ to 5° 18′.)

It is then clear that the moon's orbit will cut through the plane of the ecliptic at two points. These points, which are shown in Figure 9.3 as N and N′, are known as the *nodes* of the lunar orbit. The point N at which the moon comes from the south of the ecliptic to its north is known as the *ascending node,* and the other point is called the *descending node* (*Rāhu* and *Ketu* in Sanskrit, respecively). Naturally these points lie in opposite directions as seen from the earth at every moment of time.

As if this were not enough, there are more and more surprises all along the way.

The normal ER to the moon's orbit *does not* lie in a fixed direction. It rotates around EP (see Figure 9.3) with an angular radius of 5°.14 once in about 18.6 years. This is called the *precession of the moon's orbit.*

Box 10.2 *A demonstration of the moon's orbit and nodes*

To imagine the moon's orbit, let us take water in a trough and consider the water surface to be the plane of the ecliptic. Imagine that the earth moves around the sun in this plane. Now fix a nail or a piece of strong straight wire at the centre of a small cardboard disc in a direction perpendicular to its plane, as shown in Figure 10.4 (a). Holding the nail or wire, dip the disc in water so that its plane makes a small angle of about 5° with the surface of water (or the nail makes an angle of 5° with the vertical), until the disc is half immersed in water (see Figure 10.4 (b)).

Now one can imagine the earth to be at the centre of the disc (on the water surface or the plane of the ecliptic) and the circumference of the disc to be the orbit of the moon. The points of this orbit which just touch the water surface would be the nodes, *Rāhu* and *Ketu*.

If we now rotate the nail slowly along a circle about the vertical, always keeping its angle with the vertical equal to 5°, this would have the effect of ***Rāhu*** and ***Ketu*** revolving around the earth (always keeping diametrically opposite to each other). This is the precession of the moon's orbit in 18.6 years mentioned earlier.

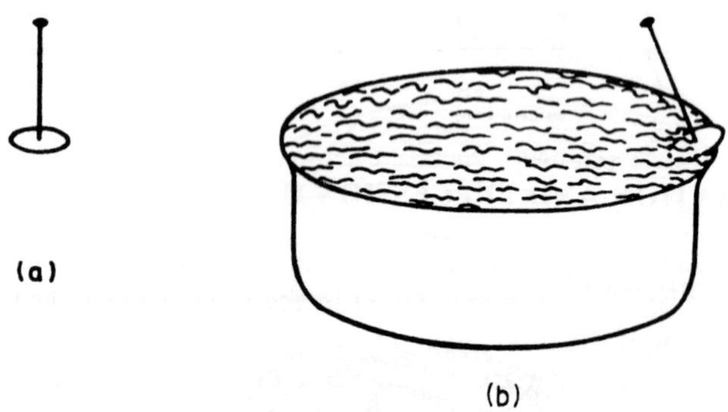

Figure 10.4 A demonstration of the combined motion of the moon around the earth and of the earth around the sun with the help of a water trough, cardboard disc and nail.

WHY ECLIPSES OCCUR

In its orbit around the earth, when the moon happens to come between the sun and the earth, it casts its shadow which falls on parts of the earth. The sun would not be visible to people in these parts on such occasions, resulting in a solar eclipse. This would naturally happen on a new moon day.

Similarly, when the earth lies between the sun and the moon, the earth's shadow may fall on the moon. A part of the moon does not receive light from the sun and hence appears dark. This causes a lunar eclipse, which would occur on a full moon day. A lunar eclipse is visible from one whole hemisphere of the earth facing the moon.

For a better understanding of the phenomena of eclipses, we must study two aspects: (a) the shadows cast by extended objects, and (b) the exact orbits of the moon around the earth and of the earth around the sun. We shall study these two aspects in that order.

SHADOWS DUE TO EXTENDED SOURCES

A mathematically ideal point source of light produces sharp shadows, as shown in Figure 10.5 (neglecting diffraction effects). However, an actual source can never be an ideal mathematical point source. Every real source has some extension. For example, an electric bulb has a luminous wire whose extension may be of the order of 2 cm, a domestic fluorescent light is about one-and-a-quarter meters long, and the sun has a laterl extension of about 1.4 million km.

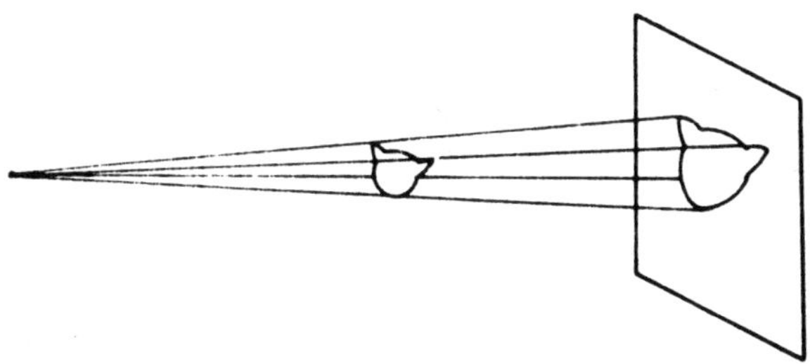

Figure 10.5 An ideal point source casts sharp shadows.

The shadows formed by such extended objects consist of various shades of bright and dark regions. The portion of the shadow which does not receive light from any point of the source is called *umbra* and is dark. The portion of the shadow which receives light from some (but not all) parts of the source is called *penumbra* and is relatively less dark. This is illustrated in Figure 10.6 for objects which are smaller or bigger than the source. Figure 10.6 (a) shows the shadow of an object CD larger than the source AB, on the screen IJ or KL. It can be readily seen that the point S between F and G does not receive light from any part of the source AB, whereas a point R between E and F or between G and H receives light only from a part of the source. Therefore FG is the umbra while EF and GH are the penumbra. The points on the screen above E or below H receive light from the entire source. One can also see what happens if the screen is moved farther from, or nearer, the object—both the umbra and the penumbra become larger and larger as the screen is moved away. The nature of the shadow remains the same.

Figure 10.6 (b) shows the shadow of an object CD smaller than the source AB on a screen IJ or KL. At the position IJ, FG is the umbra while EF and GH are the penumbra. As the screen is moved away, the umbra becomes *smaller* whereas the penumbra becomes larger. There is a critical point O beyond which the umbra vanishes altogether. The entire shadow PS in this case is penumbra because every point of the screen between P and S receives

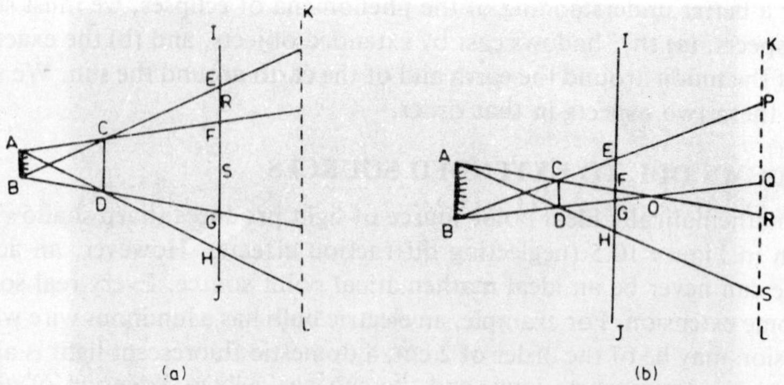

Figure 10.6 Shadows of objects due to extended sources consist of umbra and penumbra. Shadow of an object CD (a) larger than, and (b) smaller than, the source AB.

light from *some* and *only some* parts of the source. It is interesting to note that from a point between Q and R, the top and the bottom portions of the source would be seen though the central portion would be obstructed by the object CD.

This discussion has direct bearing on solar and lunar eclipses. The sun is the source of light (which is larger than both the earth and the moon), and the earth and the moon act as objects causing obstruction during lunar and solar eclipses, respectively. On a new moon day, the umbra and penumbra of the moon may sweep across the earth (Figure 10.7 (a)), obstructing the view of the sun either completely or partially for some time, depending on the position of the earth. On the other hand, if the path of the moon on a full moon day passes through the umbra of the earth as shown in Figure 10.7 (b), a total or a partial lunar eclipse results.

Notice that the earth is itself larger than the moon, its linear dimensions being about three-and-a-half times those of the moon. Therefore the earth's umbra can occasionally cover the entire moon and cause a total lunar eclipse. But the moon's umbra *never* covers the whole of the earth and therefore a total solar eclipse is never seen from *all over the earth* at the same time. Looking at it from a different perspective, this means that *a person on the moon would never see a total eclipse of the earth.*

There is one more point to be remembered. A lunar eclipse is seen only when the earth's umbra falls on the moon. If the moon passes through the earth's penumbra, it still receives light from some parts of the sun and hence its full disc is visible to us. On the other hand, a solar eclipse can occur even when the penumbra of the moon falls on the earth.

For an object smaller than the source of light, the penumbra is much larger than the umbra (see Figure 10.6 (b)). This highly increases the chances of a solar eclipse as compared to that of a lunar eclipse. On the other hand, the earth is 3.5 times bigger in its linear size than the moon, which raises the chances of a lunar eclipse. However, the first factor dominates over the second. The combined effect is that there are slightly more chances of a solar

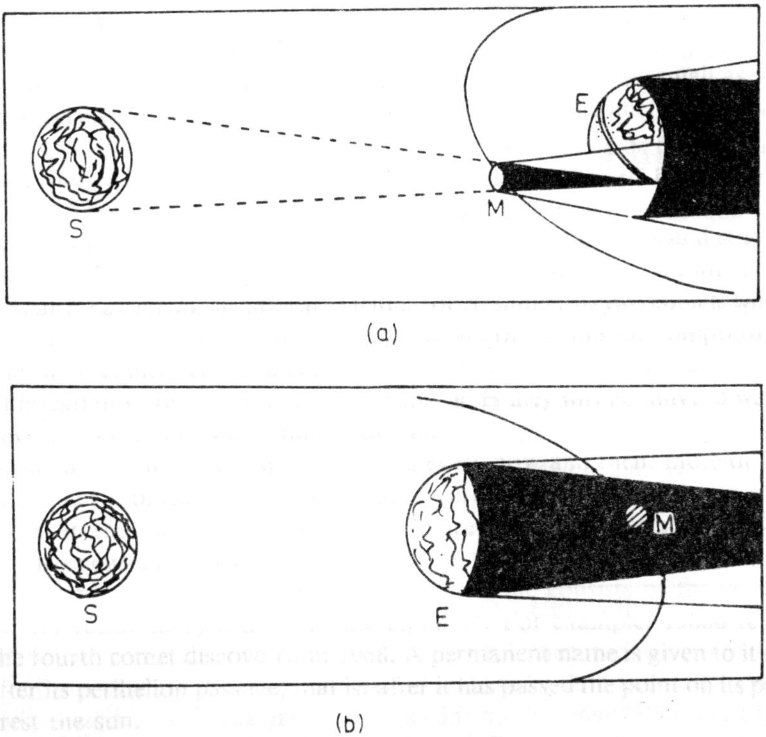

Figure 10.7 (a) During a solar eclipse, the moon casts its shadow on the earth (on a new moon day). (b) During a lunar eclipse, the moon passes through the earth's shadow (on a full moon day).

eclipse than of a lunar eclipse. You can easily verify this from the calendars of a few consecutive years (or from Table 10.1 at the end of this chapter, or from detailed geometrical calculations).

If the moon passes through the earth's penumbra, it appears to be less bright. In such a case, we talk of a *penumbral eclipse.*

You must have seen that a total eclipse of the moon does not hide it completely. Some rays of the sun grazing the earth and passing through its atmosphere are refracted and the eclipsed portion of the moon appears in a dark brown shade.

EFFECT OF MOON'S ORBIT ON ECLIPSES

But for this inclination of 5° between the earth's orbit around the sun and the moon's orbit around the earth, we would have observed a lunar eclipse on every full moon day and a solar eclipse on every new moon day. This is because a lunar eclipse takes place when the sun, the earth and the moon are in a straight line with the earth between the other two, and a solar eclipse takes place when they are in a straight line with the moon between the other two. Due to the inclination of about 5°, the moon *does not come* exactly in a straight line with the sun-earth system *on every new moon or full moon day.*

When would the moon fall in line exactly with the earth and the sun? This would occur when, during the earth's annual motion around the sun and the precession of the moon's orbit, the points *Rāhu* and *Ketu* (in addition to the moon) also fall on the sun-earth line. Thus we have two conditions for lunar and solar eclipses:

1. It should be a new moon day or a full moon day for sun-earth-moon to be *approximately* in a straight line, and

2. *Rāhu* and *Ketu* should be within a few degrees from the sun in the sky (within about 12°.25 for a total lunar eclipse and 18°.52 for a partial solar eclipse) for sun-earth-moon to be *very much* in a straight line so that the sun or the moon may eclipse the other.

Both these rules can be clearly understood if you actually perform the illustration shown in Figure 10.4.

The moon covers about 13°.2 per day (with respect to the stars). Thus one could see that on the days of lunar or solar eclipse, the moon would actually *pass through* either point *Rāhu* or *Ketu* or would be very close to either of them.

Box 10.3 *Good science and bad science*

Here is an opportunity for you to distinguish between a good scientific theory and a bad theory based on heresay or superstitions. You may have heard granny's theory about the cause of lunar and solar eclipses—that an eclipse takes place when *Rāhu* and *Ketu* gobble up the sun or the moon.

It is easy to explain yesterday's eclipse by saying that *Rāhu* or *Ketu* devoured the moon or the sun. But would you be able to predict the exact time and nature of a future eclipse by this theory? Certainly not.

To predict future eclipses, say all the eclipses which would occur the next year, one must have a thorough knowledge of the orbits of the earth and the moon, their relative inclination, the precession of the orbit of the moon in space, the points of intersection of the orbit with the plane of the ecliptic etc.

But we know that our ancestors *did* predict solar and lunar eclipses with fairly good accuracy. We know that people from all parts of the country used to throng for a dip in the holy rivers during an eclipse of which they had prior knowledge. This leads us to the thought that our ancestors must have known the scientific and technical aspects of the orbits of the earth and the moon.

Granny's theory of *Rāḥu* and *Ketu* gobbling up the sun and the moon is not the correct scientific theory of eclipses. But then we cannot expect to learn science from a common man or a man in the street. For this, we must approach knowledgeable people and apply the scientific method. When scientific and technical knowledge passes from the experts to the common man, it is bound to get distorted. Granny's theory of *Rāhu* and *Ketu* is possibly such a distortion.

FUTURE ECLIPSES

We list in Table 10.1 the solar and lunar eclipses that would occur during the decade 1991-2000. Of all the total solar eclipses in the next decade, only two will be visible over India. No total solar eclipse would be visible over India in the entire twenty-first century. In fact, for any given place on the earth, the chance of recurrence of a total solar eclipse is about once in 300 years.

Table 10.1 Solar and lunar eclipses during A.D. 1991-2000 (T: total eclipse; P: partial eclipse; A: annular eclipse)

Year	*Eclipse of the sun* (S) *or of the moon* (M)	*Date*
1991	S (A)	15-16 January
	S (T)	11 July
	M (P)	21 December
1992	S (A)	4-5 January
	M (P)	15 June
	S (T)	30 June
	M (T)	9-10 December
	S (P)	23-24 December
1993	S (P)	21 May
	M (T)	4 June
	S (P)	13 November
	M (T)	29 November
1994	S (A)	10 May
	M (P)	25 May
	S (T)	3 November
1995	M (P)	15 April
	S (A)	29 April
	S (T)*	24 October
1996	M (T)	3-4 April
	S (P)	17 April
	M (T)	27 September
	S (P)	12 October
1997	S (T)	8-9 March
	M (P)	24 March
	S (P)	1-2 September
	M (T)	16 September
1998	S (T)	26 February
	S (A)	21-22 August
1999	S (A)	16 February
	M (P)	28 July
	S (T)*	11 August
2000	M (T)	21 January
	S (P)	5 February
	S (P)	1 July
	M (T)	16 July
	S (P)	31 July
	S (P)	25 December

*These total solar eclipses would be seen over some parts of India.

11

Planetary Orbits and Kepler's Laws

Before studying all the other major planets, minor planets, their satellites, comets and such other bodies in the solar system separately, it would be proper to discuss their common features. The motion of all these bodies is essentially decided by the all-pervading gravitational attraction which exists between any two bodies.

In this chapter, we shall study the true orbit of a planet around a massive sun or of a satellite around the parent planet. The next chapter is devoted to some astronomical and physical considerations common to all planets.

Kepler formulated the first systematic laws of planetary motion in the beginning of the seventeenth century primarily on the basis of the observations of Tycho Brahe. This was followed by Galileo's first observations through a telescope. It culminated in Newton's theory of universal gravitation. This laid a strong mathematical foundation and all the Keplerian laws of planetary motion were derived on the basis of Newton's theory. We shall now discuss the real nature of planetary orbits followed by Kepler's laws.

What are the orbits of the planets really like? Are they circular as we are told in junior school?

The paths of planets are not circular. They are approximately elliptical. An *ellipse* is like a flattened circle, somewhat similar to the shape of an egg.

HOW TO DRAW AN ELLIPSE

It is posssible to draw an accurate ellipse relatively easily. For this, let us fix two points F and G on a sheet of paper, as shown in Figure 11.1. Take a thread or string and tie it into a loop which should enclose the two pins. Take a pencil, and with its help, stretch the loop into a triangle FPGF, with

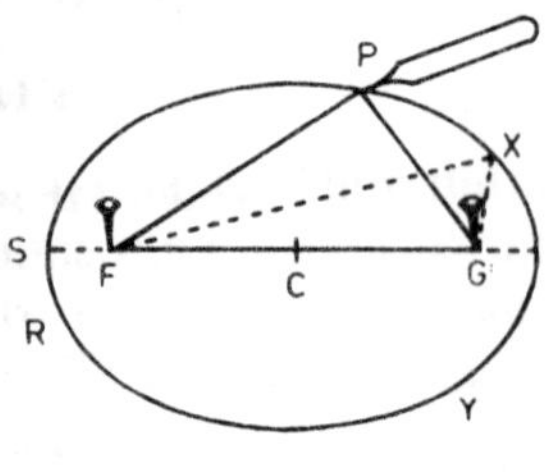

Figure 11.1 Drawing an ellipse.

portions of the taut string for the sides of the triangle. Keeping the loop always stretched, trace a curve PXTYRSP on the paper with the pencil. This curve is an ellipse.

The two points F and G (the positions of the pins) are called the two *foci* of the ellipse.

An ellipse thus has the property that for any point X on it, the sum of its distances from the two foci, FX + XG, is a constant. The shape and size of an ellipse are totally determined by two quantities, the distance between the pins and the length of the thread. We can get ellipses of different shapes and sizes by changing the distance FG between the pins and the length of the loop. Thus, we can get an extremely flattened ellipse by increasing the distance FG, keeping the length of the loop the same. On the other hand, if we bring the two foci closer to each other, we get an ellipse which looks almost like a circle. Two such ellipses are shown in Figure 11.2.

Finally if the two pins are brought together, that is, we keep only one pin, and draw the curve as above, it would be just a circle. All points of this circle would be equidistant from the single pin, which would then be said to be its centre. Thus we see that a circle is a special case of an ellipse.

Box 11.1 *Axes and eccentricity of an ellipse*

An ellipse is characterized by an important parameter known as its *eccentricity*. It gives a quantitative, mathematical measure for the deviation of the ellipse from a circle. To understand this, consider the point C in Figure 11.1 or Figure 11.2, which is the midpoint of the line joining the two foci F and G. We extend the line FCG on both sides to meet the ellipse at points S and T. Then the eccentricity of the ellipse is defined as the ratio of the distance CG (distance from the centre to either focus) to CT (largest distance of any point on the ellipse from the centre), that is

$$e = CG/CT = CF/CS.$$

S and T are the farthest points of the ellipse from the centre C. ST is called the *major axis* of the ellipse. Half of it, that is CT, is called the *semimajor axis* and is generally denoted by the symbol a. Let us draw a line perpendicular to the major axis in the plane of the ellipse and let it intersect the ellipse at the points X and Y as shown in Figure 11.2. Then X and Y are the points on the ellipse nearest to the centre. XY is called the *minor axis* and half of it, that is CX, is called the *semiminor axis*. It is generally denoted by b. Then the eccentricity of the ellipse can also be related to the semimajor and the semiminor axes, and is given by

$$e = (a^2 - b^2)^{1/2}/a .$$

The eccentricity of an ellipse always lies between zero and unity. The ellipse of Figure 11.2 (a) has a larger value of e, equal to 0.8, while that of Figure 11.2 (b) has a smaller e, equal to 0.3. Both of them have the same major axis. It should be clear that a circle is an ellipse with $e = 0$.

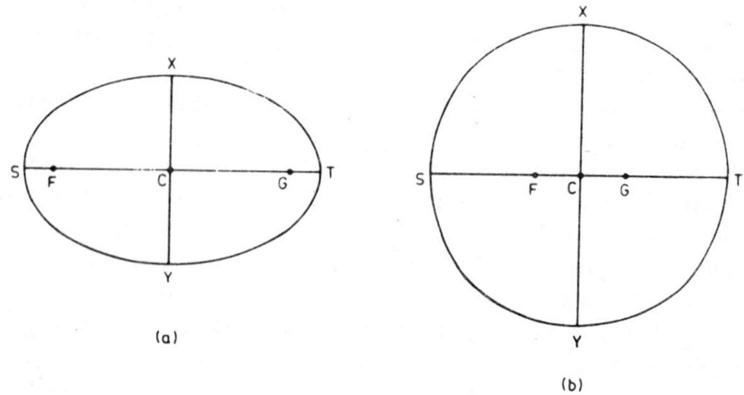

Figure 11.2 Ellipses of different shapes.

HISTORICAL DEVELOPMENT IN UNDERSTANDING PLANETARY ORBITS

Until the middle ages, it was thought that the earth is at the centre of the entire universe and all the stars and planets, including the sun are moving around it in circles. The view changed around the fifteenth century and scientists generally accepted that the earth is not at the centre of the universe. On the other hand, the planets revolve around the sun in circular orbits. In the beginning of the seventeenth century, Kepler suggested that the paths of planets are not circular but elliptical, with the sun at one of the foci, as in Figure 11.3.

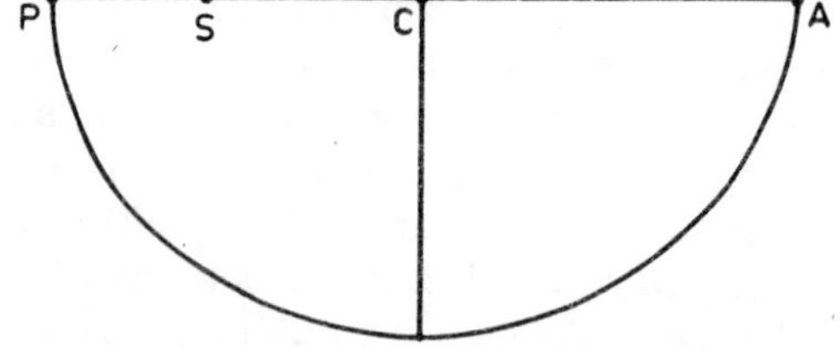

Figure 11.3 A planet going round the sun S in an elliptical orbit. P is the perihelion and A the aphelion, C is the centre of the ellipse and CA is the semimajor axis.

There is a reason why the theories of circular orbits succeeded to a great extent. The maximum eccentricity of all the planetary orbits, excluding Pluto's, is that of Mercury, which has $e = 0.206$. Now for this value of e, the semiminor and the semimajor axes are related by $b = 0.978a$, that is, *b* is only about 2% smaller than *a* (a difference of 1 mm in a circle of 5 cm radius). Such an ellipse will obviously appear as a circle even to the best human eye. The only difference is that the sun will not be at the centre of the approximate circle but about $0.2a$ (20% of the semimajor axis) away from the centre.

During its revolution around the sun, a planet's distance from the sun thus constantly changes. The point P on its path (Figure 11.3) which is nearest to the sun is called the *perihelion* and the point A which is farthest from the sun is called the *aphelion*. The deviation of the ellipse from circular orbits is noticeable only for some planets (for example, Mercury and Pluto). In the case of other planets, the orbits are almost nearly circular (see Table 11.1).

However, there are certain objects, all the comets and some asteroids, for example, which go around the sun in highly elliptical orbits. Halley's comet, which approached the sun latest in January 1986 and which has a period of about 76 years, is one of the best known comets. Its orbit is highly elliptical ($e = 0.967$).

Table 11.1 Eccentricity of the paths of some objects in the solar system

Class	*Object*	*Eccentricity*	*Ratio* SA/SP	*Ratio b/a*
Planet	Pluto	0.247	1.656	0.967
	Mercury	0.206	1.519	0.979
	Mars	0.093	1.205	0.9957
	Earth	0.017	1.035	0.999 86
	Venus	0.007	1.014	0.999 975
Satellite	Moon (of Earth)	0.0549	1.116	0.998 49
Asteroid	Icarus	0.827	10.56	0.562
	Adonis	0.78	8.09	0.626
	Hidalgo	0.65	4.71	0.661
Comet	Hirschel Rigolet	0.974	75.9	0.227
	Halley	0.967	59.6	0.255
	Encke	0.847	12.1	0.532

For the sake of interest, we give in Table 11.1 the values of eccentricity of the orbits of some planets, comets, minor planets (asteroids), and the moon. The ratio of aphelion to perihelion distances (that is, of the largest distance to the smallest distance from the sun or the earth in the case of the moon, that is SA/SP in Figure 11.3) is also given in each case. You will notice that the orbit of Venus is almost circular, with only 0.0025% difference (1 mm in 40 metres) between the major and minor axes, but the difference between its largest and smallest distances from the sun is about 1.4%. Among the planets, Pluto has the most elliptical orbit. Its maximum distance from the sun is about 66% larger than its minimum distance. The orbits of asteroids and comets are highly elliptical. The farthest distance of the comet Hirschel Rigolet is about 76 times its least distance from the sun.

PERTURBATION DUE TO OTHER PLANETS

Further discrepancies were observed as early as the seventeenth century in the elliptical paths of planets. It was noticed that after each revolution, a planet does not trace the same ellipse over and over again, but appears to have shifted slightly. It thus traces a new ellipse in each revolution. This is shown in Figure 11.4 in a rather exaggerated manner. P, P′ and P″ in this figure are the positions of the perihelion of a planet in successive orbits. This effect is known as the *advance of the perihelion of a planet*. We express this by saying that the path of a planet is *not a closed ellipse*.

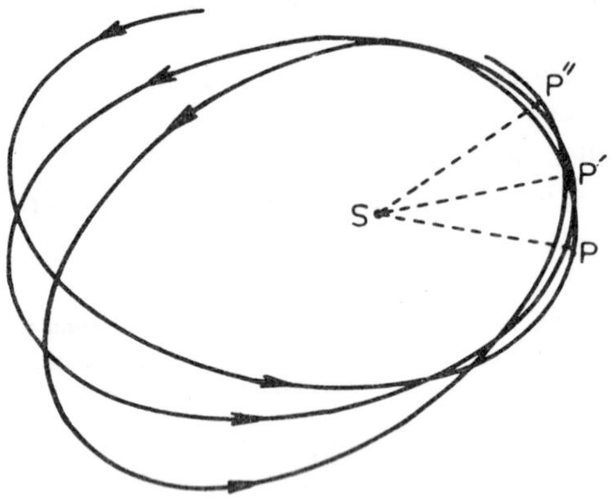

Figure 11.4 Advance of the perihelion of a planet (highly exaggerated).

Actually this shift, or the rotation of the line SP connecting the sun to the perihelion P of a planet, is extremely small. Among the planets, its largest value occurs for Saturn. The values for a few planets are shown in Table 11.2. The perihelion of Saturn advances by 2020″ per century and that of Mercury by 574″ per century, which comes to about 9 minutes 34 seconds of arc (less than ten arcmin, or less than one-sixth of a degree; see Box 4.1). The perihelion of Mercury would thus advance by less than 1° in six centuries. It is indeed highly creditable for the astronomers of the seventeenth century to have made quite accurate measurements of the positions of planets.

Table 11.2 Advance of perihelion of planets

Planet	*Advance of perihelion (in arcsec per century)*
1. Saturn	2020
2. Mars	1600
3. Earth	1150
4. Jupiter	770
5. Mercury	574

One explanation for this phenomenon was immediately provided by astronomers. The solar system consists of several planets, and each planet must be affected by the combined gravitational pull of the remaining planets, in addition to the sun, causing a slight deviation from the strictly elliptical path of a planet. Rigorous calculations were therefore made to find out these effects on the planets. It turned out that these calculations gave exceedingly good agreement for all planets *except* Mercury. For Mercury, calculations gave a value of 531″ per century for the advance of the perihelion of Mercury, which was still short of the measured value by about 43″ per century.

Physicists tried to give several explanations to account for this discrepancy, but all such theories failed in one respect or the other.

TEST OF GENERAL RELATIVITY

Then in 1915, Einstein developed the *general theory of relativity*. Conceptually, this theory is a radical modification of Newton's theory of gravitation and differs from the latter in predicting some very new, subtle effects. One of the consequences of the general theory of relativity is that the path of a planet around a star must be an ellipse whose axis would rotate with time even if there is no third body to perturb it. When the mass of the sun and the velocity and the distance of Mercury from the sun are substituted in Einstein's formula, it gives the correct value of 43″ per century for the precession of the orbit of Mercury. This provided one of the first three tests of Einstein's general theory of relativity right at the moment of its inception.

Box 11.2 *If Mercury were the only planet of the sun*

If Mercury were the only planet around the sun, its orbit should be a closed ellipse according to Newton's theory of gravitation, while it must be an ellipse whose perihelion is advancing at the rate of 43 seconds of arc per century according to Einstein's general theory of relativity. It further precesses through 531″ per century due to perturbations of other planets, making a total of 574″ per century.

MODIFICATION DUE TO EXISTENCE OF SATELLITES

As if this were not enough, further minute discrepancies were observed in the motion of the moon and the earth. It was found that the moon was not traversing an exact elliptical path and that the earth was wobbling in its path around the sun.

If a planet has no satellite, it would approximately describe an ellipse around the sun (neglecting the advance of the perihelion just described). In the case of a planet with satellites, it is the *centre of mass* of the planet and its satellites that would move along the ellipse:

We have already explained in Chapter 10 (see Figure 10.3) that the earth would follow a wiggly path around the sun. This effect occurs due to the presence of the moon. Since the earth takes a year to go round the sun and the moon takes about 29 days to go round the earth (with respect to the sun), the earth would wobble 12 to 13 times about its path in one revolution.

Since the earth has only one satellite, even this wobbling is fairly systematic. In the case of planets with many satellites, such as Jupiter and Saturn, this would become rather complicated. The centre of mass of a planet with many satellites would depend on the positions of the satellites. It would therefore change its position from time to time. Consequently, the position of such a planet would seem to wobble erratically about its mean elliptical path. However, in the case of all planets with more than one satellites—Mars, Jupiter, Saturn, Uranus and Neptune—the mass of the parent planet is very must larger than the combined mass of its satellites. Hence the centre of mass of the planet-satellites system in these cases is quite close to the centre of the planet and the wobble is quite small.

Such are the ways of nature. More and more discrepancies are discovered through more and more accurate modern methods of measurement. These give rise to new effects and new theories and the chain continues.

KEPLER'S LAWS OF PLANETARY MOTION

The laws formulated by Kepler are valid for the motion of a planet around a star or for a satellite (natural or artificial) around a planet, and take into account their elliptical orbits.

Kepler's laws of planetary motion are:

1. Each planet moves around the sun in an elliptical orbit with the sun at one focus. (Actually, the sun and the planet both revolve around their common centre of mass in elliptical orbits.)

2. The velocity of a planet in its path varies in such a way that the line connecting the sun with the planet sweeps out equal areas in equal time intervals.

3. The cube of the mean distance of the planet from the sun is proportional to the square of the period of revolution of the planet around the sun.

The first law (and the slight departure from it) has already been explained extensively in the last few pages. To understand Kepler's second law, consider again a planet going around a star, moving in an elliptical orbit, as in Figure 11.5. Let P_1, P_2, P_3, and P_4 be the successive positions of the planet

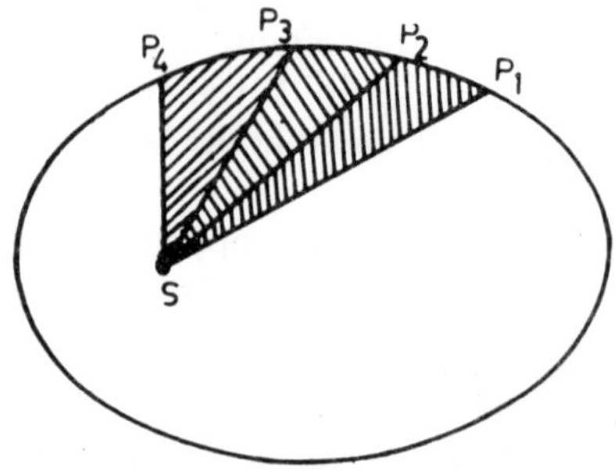

Figure 11.5 Explanation of the Kepler's second law (schematic, not to scale).

at times t_1, t_2, t_3, and t_4, such that the time intervals between these positions are equal, that is $t_4 - t_3 = t_3 - t_2 = t_2 - t_1$. Then the second law states that the areas P_1SP_2, P_2SP_3 and P_3SP_4 are equal to each other.

Since the distance of the planet from the star is changing (for example, SP_1 is larger than SP_2, which in turn is larger than SP_3, etc), it follows that at smaller distances from the star, the planet must be travelling longer paths than at larger distances (that is, the path P_3P_4 is longer than P_2P_3 which is itself longer than P_1P_2). This is a consequence of the observed fact that the speed of a planet at its perihelion is the largest, while at the aphelion it is the least. (Note that in the case of a circular orbit, the linear speed of a planet at all points on its path is a constant.)

Kepler's third law is useful for comparing the periods of revolution of the various planets around a common star. If the mean distances of the planets from the star are a_1, a_2, a_3, ..., a_n, and their periods of revolution are P_1, P_2, P_3, ..., P_n, then this law states that

$$\frac{a_1^3}{P_1^2} = \frac{a_2^3}{P_2^2} = \frac{a_3^3}{P_3^2} = \ldots = \frac{a_n^3}{P_n^2}.$$

Box 11.3 *An example of Kepler's third law*

Using the data for the earth and the fact that Jupiter completes one revolution around the sun in 11.9 years, we can find the mean distance of Jupiter from the sun.

If a_E, P_E are the relevant parameters for the earth and a_J, P_J those for Jupiter, then according to Kepler's third law,

$$\frac{a_E^3}{P_E^2} = \frac{a_J^3}{P_J^2},$$

or

$$a_J^3 = (P_J^2/P_E^2)\, a_E^3.$$

The distance of the earth from the sun is $a_E = 1.5 \times 10^{11}$ m, and its period of revolution is 1.00 year. Using these values in the above equation, we have

$$a_J^3 = [(11.9)^2/(1.00)^2]\,(1.5 \times 10^{11})^3 \text{ m}^3 = 4.78 \times 10^{35} \text{ m}^3.$$

This gives

$$a_J = 7.82 \times 10^{11} \text{ m}.$$

Thus Jupiter is slightly more than five times farther away from the sun than is the earth.

12

Common Features in the Study of Planets

There are several common features about planets (and also about satellites and comets) which should be known in order to have some idea of the solar system. They are both of astronomical and physical nature.

The astronomical parameters include the concepts of conjunction, opposition and transit, the revolution of the planet around the sun and the nature of its orbit, nodes etc. The physical parameters include the rotation of a planet about itself, the axis of its rotation, its mass, density and gravity on its surface, its brightness, its contents, atmosphere, temperature, magnetic field, etc.

The planets are not self-luminous. They are visible to us because they partially reflect sunlight falling on their surface (see Box 2.1). The ratio of light which a non-luminous body reflects to that which it receives is called its *albedo*. This is one of the important physical parameters because it has an effect on the brightness and the temperature of the object. We shall see later that Venus has the highest albedo of all the bodies in the solar system.

All the concepts developed in this chapter apply equally well to satellites of planets, asteroids, comets, etc.

CONJUNCTION AND OPPOSITION

Those planets whose orbits lie inside the orbit of the earth are called *inferior planets* and those whose orbits are outside that of the earth are called *superior planets*. Figure 12.1 shows the orbits of the earth, an inferior planet, and a superior planet around the sun. For the sake of simplicity, we have shown all the three orbits to be circular and in the same plane, although they are not really so.

An inferior planet is in line with the sun and the earth at the two positions A and B shown in Figure 12.1. The planet is said to be in *conjunction* with the sun at these positions. When it is between the sun and the earth, such as at the point A, it is said to be in *inferior conjunction* with the sun. At the other point B, where it hides behind the sun, it is said to be in *superior conjunction* with the sun. The word 'conjunction' is used to describe this phenomenon because the planet appears to conjoin the sun.

Let ER and ET be the tangents to the orbit of the inferior planet from the position of the earth. It is evident that the maximum angular separation of the planet from the sun in the sky is the $\angle SER$ or $\angle SET$. This is called the *greatest solar elongation* of the planet. The point R is the point of *eastern*

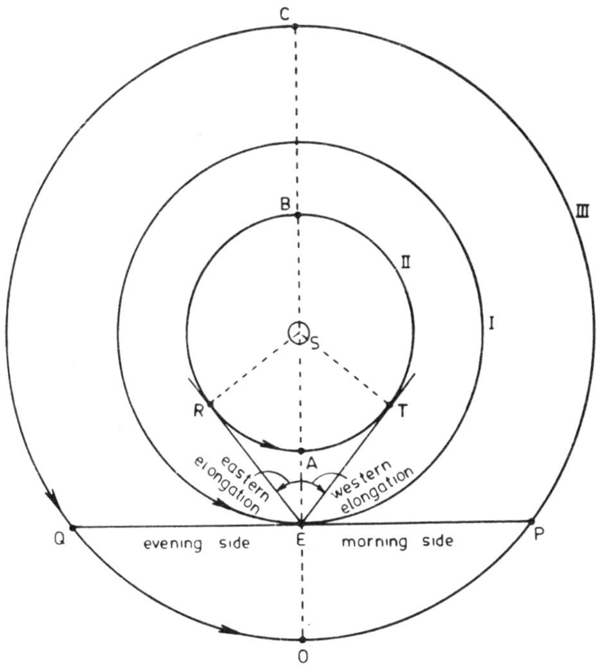

Figure 12.1 The orbits of the earth (I), an inferior planet (II) and a superior planet (III) around the sun (S).

solar elongation while T is the point of western solar elongation. We have already seen that the greatest solar elongation of Mercury is 28° while that of Venus is 47°. You can thus verify that the maximum distance of Mercury from the sun is sin 28° AU or 0.47 AU, while that of Venus is sin 47° AU or 0.73 AU. These are the only planets in our solar system whose orbits lie inside that of the earth.

Let QP be the normal to the line ES passing through the earth and intersecting the orbit of a superior planet at points Q and P. When the outer planet is at either of these positions, it is 90° away from the sun in the sky. It is therefore said to be in *solar quadrature* at these points on its path. Thus if the sun is in the constellation Taurus, a planet in quadrature at Q and P must be in Leo and Capricorn, respectively.

An infereior planet can never be in quadrature. The maximum elongation of an inferior planet must be less than 90°.

Let the line joining the earth to the sun intersect the orbit of the outer planet at points O and C, as shown in Figure 12.1. The planet is said to be in *conjunction* at C and in *opposition* at O.

Thus an inferior planet would be nearest the earth at inferior conjunction, while a superior planet would be nearest at opposition. The shortest distance of a planet from the earth is not a constant; it changes at each revolution due to the fact that the orbits of the planet and the earth are elliptical and they do not lie in the same plane.

A planet or star to the right of the sun S in Figure 12.1 would be seen in the morning sky before sunrise, and one to the left of S would be seen in the evening sky after sunset.

When an inferior planet comes exactly between the earth and the sun at inferior conjunction, it would appear to pass *over* the disc of the sun. Such a phenomenon is known as a *transit* and can be observed with the help of a telescope of moderate power. During a transit, Mercury or Venus appears as a tiny black dot crossing the disc of the sun. On such an occasion, the apparent diameter of Venus is about one-thirtieth of that of the sun while that of Mercury is even smaller. Figure 12.2 shows a photograph of Mercury in transit across the sun.

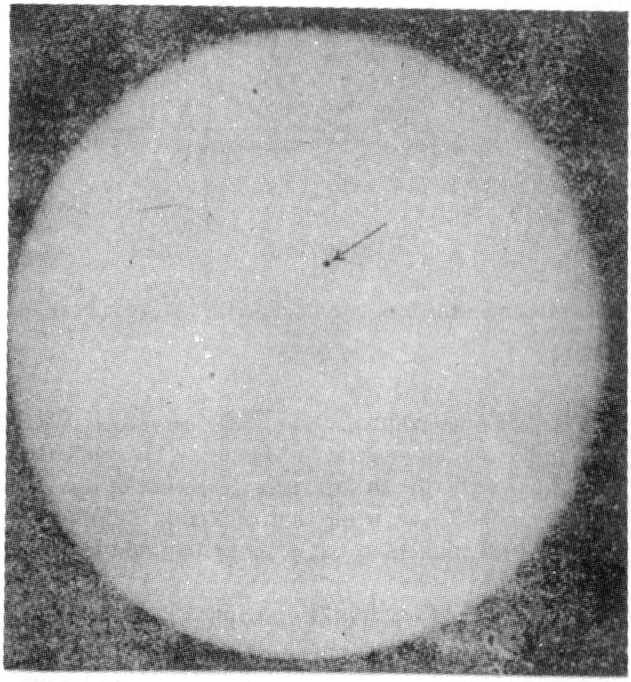

Figure 12.2 Mercury in transit across the sun. (Photographed on November 10, 1973, at *Vedhashālā,* Ahmedabad. *Courtesy*: PRL)

Superior planets also show phases, though not all. Near conjunction and opposition (points C and O in Figure 12.1), a superior planet turns its fully illuminated face towards us. It is at the point O that a superior planet comes closest to the earth. Since Mars is the nearest superior planet to the earth and the eccentricity of its orbit is appreciable, it does not appear to be equally bright at all its oppositions. During its most favourable oppositions it comes as close as 0.37 AU and at the least favourable oppositions it is as far as 0.67 AU from the earth.

Box 12.1 *How common are the transits of planets?*

Since the orbits of the planets are not in the same plane, the transit of an inferior planet across the sun does not take place in every revolution. In fact, the conditions for this to happen are the same as those for the moon to eclipse the sun. The transits of Mercury occur at the rate of roughly 13 per century. The transits of Venus are much rarer. The last transit of Venus took place in December 1882. The dates of some future transits of Mercury and Venus are given in Table 12.1.

Table 12.1 Some future transits of Mercury and Venus

Mercury	*Venus*
November 5, 1993	June 8, 2004
November 15, 1999	June 6, 2012
May 6, 2003	December 10, 2117
	December 8, 2125

Figure 12.3 shows how a superior planet shows its maximum phase at quadrature. The angle AQC or angle BQD in this figure is the maximum *gibbous phase* of the superior planet.

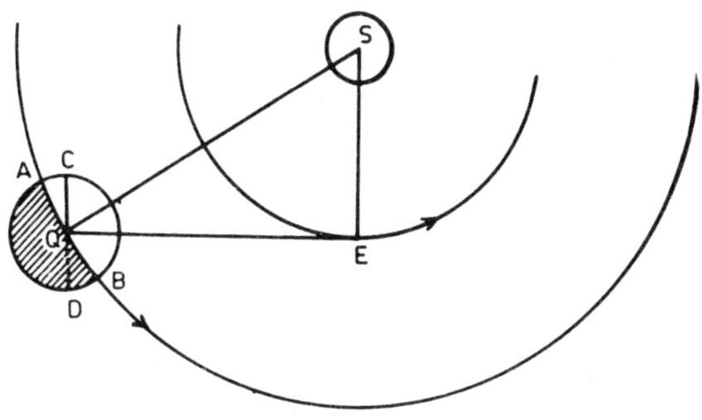

Figure 12.3 A superior planet shows its maximum gibbous phase when it is at quadrature.

THE BRIGHTNESS SCALE

The brightness of a star depends on how much energy it emits per second. In the case of a planet, satellite or a comet, it is a measure of how much solar radiation it reflects.

However, two stars which are equally bright may not *appear* to us to be so because of their different distances from the earth. A scale to measure the *apparent brightness* of a stellar object has been introduced by astronomers. It is known as the *apparent magnitude* of an object. It is a peculiar, logarithmic scale.

Brighter objects have a smaller index on this scale while fainter objects have a higher index. The index can be positive as well as negative. Thus an object with apparent magnitude −1 is brighter than one with magnitude 0, which in turn is brighter than an object with magnitude +1.

A unit change in magnitude corresponds to a brightness ratio of 2.512, so that a difference of 5 units in the apparent magnitude corresponds to a brightness ratio of $(2.512)^5 = 100$. Table 12.2 lists the apparent magnitudes of some members of the solar system. An object having an index 0 on this scale would be 100 times brighter than one having an index +5, which again would be 100 times brighter than one having an index +10, and so on.

The limit of vision comes approximately at an apparent magnitude of +6. This means that stellar objects with an apparent magnitude of less than +6 would be visible to the naked eye while those with an index greater than +6 would be invisible.

The sun obviously has the least value of apparent magnitude among all the stellar objects. Uranus has an apparent magnitude of +5.5, which is near the limit of the naked-eye vision, and can be seen with great difficulty only if its position is known beforehand.

Table 12.2 The apparent magnitudes of some members of the solar system

Stellar object	*Apparent magnitude (planets at their mean opposition)*
1. The sun	−26.7
2. Full Moon	−12.7
3. Venus	−4.4
4. Jupiter	−2.6
5. Mars	−1.9
6. Saturn	+0.7
7. Uranus	+5.5
8. Ceres (asteroid)	+6.9
9. Neptune	+7.6
10. Pluto	+14.8

PARAMETERS OF PLANETARY ORBIT

What are the parameters which specify the orbit of a planet completely? What information must be looked for and mentioned in characterizing a planetary orbit with respect to the earth? We shall see that the following seven parameters give complete information about the orbit of a planet, although only six of them are really necessary. These are called the *orbital elements* of a planet.

1. One of the most important parameters for a planet is the angle between its plane of revolution and the earth's plane of revolution around the sun.

(We have mentioned earlier that the latter is called the plane of the ecliptic.) This angle is denoted by the symbol *i* and known as the *angle of inclination*. No two planets revolve around the sun in the same plane. The solar system is thus a three-dimensional structure rather than a two-dimensional one.

Figure 12.4 shows the plane of the ecliptic and the plane of a planet's orbit. The angle between these two planets is the angle *i* we talked about. The orbit of the planet intersects the plane of the ecliptic at points N and N′. These are called the *nodes*. The point N at which the planet comes from the south of the ecliptic to its north is known as the *ascending node*. The other node N′ where it crosses the ecliptic from its northern side to the southern side is called the *descending node*. You will notice that this is very similar to what we have seen in connection with the moon in Chapter 10.

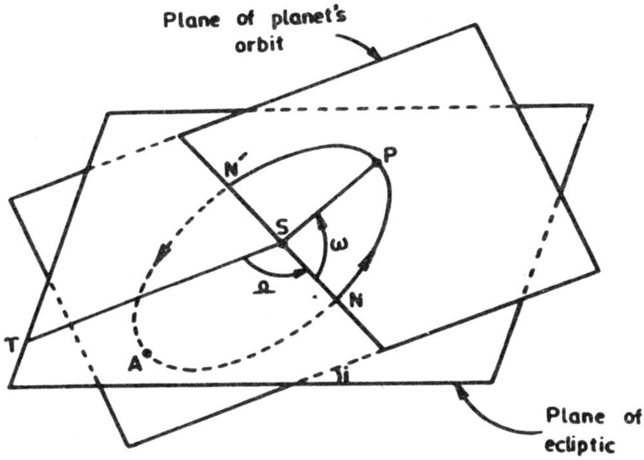

Figure 12.4 The orbital elements of a planet. S is the sun, ♈ denotes the vernal equinox, P the perihelion and A the aphelion. Other parameters are explained in the text.

2. We must have a parameter to specify the direction of the line of orbital nodes. For this we take the vernal equinox as the reference point and measure the angle of the line SN from the line S♈. The angle ∠♈SN is known as the *longitude of the ascending node* and is denoted by Ω. Note that the line SN is common to both the planes while the line S♈ lies in the plane of the ecliptic. The angle Ω is therefore to be measured in the latter plane.

These two parameters, *i* and Ω, give us the location of the plane of the planetary orbit in space.

3. The next piece of information is the length of the semimajor axis *a* of the elliptical orbit. This can also be taken as the average distance of the planet from the sun (a time-average over one period of revolution). It gives us an idea of the size of the orbit.

4. The fourth parameter is the *eccentricity e* of the elliptical orbit. This gives an idea of the shape of the elliptical orbit.

5. We need a parameter to specify the orientation of the ellipse in the plane of motion of the planet. This is done by giving the direction of the perihelion

of the planet with respect to the node N. Therefore our next parameter is the angle ∠NSP, which is the angle from the ascending node to the perihelion. It is known as the *argument of the perihelion* and is denoted by ω. Note that the line SP lies in the plane of the planetary orbit, and hence ω is to be measured in this plane.

The first five parameters specify the geometry of the planetary orbit in space. There are two more parameters to specify the dynamical motion of the planet in this orbit.

6. The next parameter is the time taken by the planet to complete one revolution around the sun (with respect to the stars). This is known as the *sidereal period of revolution* and is denoted by P. Note that this is not independent of the length of the semimajor axis a. In fact, we can derive one from the other using Kepler's third law.

7. We must know the date of the last passage of the planet through its perihelion, the point P in Figure 12.4. This will enable us to find out its exact position at any moment. This is denoted by T and it also known as the *epoch* of the planet.

All these parameters are summarised in Box 12.2. Only six of the seven parameters are independent.

Box 12.2 *The orbital elements of a planet*

The seven orbital elements are summarised below. The geometrical information they give is also mentioned.

Location of the plane of the orbit in space: angle of inclination, i, and longitude of the ascending node, Ω.

Orientation of the ellipse: angle of the perihelion from the ascending node, ω.

Size of the elliptical orbit: semimajor axis (or average distance of the planet from the sun), a.

Form of the ellipse: eccentricity of the ellipse, e.

Position of the planet at any time: period of revolution, P, and date of the last perihelion passage, T.

The orbit of a planet is not a static thing. All the above seven parameters change with time, although very slowly for planets. There are several causes responsible for changes in these parameters. Some of these are: (a) perturbations due to other planets, (b) the general relativistic effect of the advance of the perihelion of a planet (see Chapter 11), (c) the precession of the orbit (see Chapters 6 and 10 for the cases of the earth and the moon, respectively), (d) precession of the equinoxes (see Box 7.1) and so on. There are several other causes of more or less importance which have not been introduced in this book.

13

The Inferior Planets—Mercury and Venus

Mercury and Venus are the two planets closer to the sun than is the earth. They have been the subject of much attention during the last few decades when many spacecraft were launched towards them. We shall now study them.

MERCURY

Mercury is the second smallest planet of the solar system and is bigger only than Pluto. It is also very close to the the sun in the sky. We have seen that its maximum elongation is 28°. It can therefore be observed with great difficulty. At its maximum elongation, it is above the horizon for less than two hours before sunrise or after sunset. However, we must leave aside a considerable period of time for twilight during which all the objects in the sky except the sun and the moon remain invisible. This leaves us with a short period suitable for observing Mercury, about an hour before daybreak or after nightfall, when it is at its maximum favourable elongation. It cannot be observed during other times of the year.

Fortunately in India, the sky is clear most of the year. It can therefore be seen relatively easily before daybreak or after nightfall, sometimes up to one-and-a-half hours before sunrise or after sunset. It had been observed by our ancestors several centuries ago. With large telescopes, however, Mercury can be observed throughout the day in spite of the bright sun.

Colour Plate 5 shows Mercury, photographed by Mariner 10 spacecraft from a distance of 200 000 km. This spacecraft passed near Mercury thrice, in March 1974, September 1974, and March 1975.

ASTRONOMICAL PARAMETERS OF MERCURY

The orbit of Mercury is inclined at 7° to the plane of the ecliptic. Its path around the sun is highly elliptical. The eccentricity of its orbit is 0.206, which is next to Pluto's among the major planets of the solar system.

The period of revolution of Mercury around the sun is 88 days. The average semimajor axis of its orbit is 57.9×10^6 km (about 58 million km). Because of the high ellipticity of the orbit, its closest distance to the sun is 45.9×10^6 km. At its farthest, it is 69.8×10^6 km away from the sun. The aphelion distance is thus 52% higher than the perihelion distance. Its mean speed in its orbit is 47.9 km/s with respct to the sun.

The earth comes in line with the nodes of Mercury in May and November. If Mercury is near the corresponding node at this time, a transit of Mercury is seen across the sun (see Figure 12.2 and Table 12.1).

PHYSICAL PARAMETERS OF MERCURY

Mercury has a radius of 2430 km. Since its distance from the earth varies from about 100 million km to about 200 million km, its angular diameter varies between 5″ and 13″ (5 and 13 seconds of arc).

Since Mercury has no satellite, its mass is very difficult to calculate. It had to be calculated from the effect of Mercury on comets when they pass near it. As a comet passes near a massive planet, its path is perturbed. Such perturbations were observed in the paths of several comets which flew near Mercury. The earlier data in this regard were not reliable and therefore the correct mass of Mercury was not known for a long time. With more accurate methods of observation based on artificial satellites and spacecraft, a more reliable value for it has been obtained. It is now believed that the mass of Mercury is 5.5% of the mass of the earth, or 3.3×10^{23} kg.

The period of rotation of Mercury about itself is about two-thirds of the period of revolution around the sun. This simple ratio 2:3 of its period of rotation to its period of revolution has been given the name *spin-orbit resonance*. The fact that this ratio is simple may not be a mere coincidence. Due to this combination of the two periods, it has long days and nights, each lasting about three earth-months.

Owing to its proximity to the sun and long days, its surface temperature during the day is of the order of 430°C, which is sufficient to melt lead. The temperature on the dark side is believed to be very low, about −180°C.

Mercury has no atmosphere. Its outline therefore appears to be very sharp.

Its surface reflects only about 6% of the light received from the sun. This means that Mercury has a very small albedo. This is another reason why Mercury appears to be very faint. Mariner photographs show that the surface of Mercury is cratered like that of the moon.

Colour Plate 6 shows a close-up of Mercury.

VENUS

Venus is the second planet from the sun. Its orbit is closest to that of the earth among all the planets.

ASTRONOMICAL PARAMETERS OF VENUS

Venus completes a revolution around the sun in 224.7 days. The average radius of its orbit is 108 million km. Its orbit has the least ellipticity of all the planets. Its farthest distance from the sun is therefore only 1.4% larger than its minimum distance. Its orbit is inclined to the plane of the ecliptic at an angle of 3°.4. Its mean speed around the sun is 35 km/s with respect to the sun.

On favourable occasions its least distance from the earth is only 40 million km. On such occasions, its visible diameter grows to 63″. At its farthest (most unfavourable superior conjunction), its visible diameter shrinks to 10″.

The transits of Venus across the sun are very rare. They occur only in June or December, since this is when the earth comes in line with the nodes of Venus. If Venus happens to be at its node nearer the earth, a transit can be seen. Venus takes about 8 hours to cross the sun's disc under optimum conditions. There shall be no transit of Venus in the twentieth century. (See Table 12.1 for future transits of Venus.)

PHYSICAL PARAMETERS OF VENUS

The radius of Venus is 6100 km, which includes a thick cloud layer. This is only slightly smaller than that of the earth. Its albedo (reflectivity of its surface) is 72%, which is the highest among all the celestial bodies. It therefore appears very bright in the sky. In fact, it is the third brightest object in the sky after the sun and the moon. On a dark, moonless night, it casts visible shadows of objects on the earth. During its brightest phases, it can easily be spotted in the daylight.

It rotates about its own axis in 243 days. This is the slowest period of rotation among all planets. In fact, it is longer than its period of revolution around the sun. This rotation is retrograde, so that the sun appears to rise in the west and set in the east. The length of the day on Venus is about 117 days.

As it has no satellite, its mass also had to be calculated from cometary perturbations and from (artificial) satellite measurements. A very accurate value was obtained from the data of Pioneer Venus Orbiter which was put into orbit around Venus. The mass of Venus is now believed to be about 81.5% of that of the earth, or 4.869×10^{24} kg. In size and mass, Venus is almost like a twin of the earth among all the planets.

It has very thick atmosophere, mostly consisting of clouds of carbon dioxide and small proportions of many other gases. Due to this, it has not been possible to get much information about its surface from earth-based telescopes or even from space probes. It is believed that the atmospheric pressure on its surface is about a hundred times that on the earth's surface. Very dim sunlight can penetrate through such thick and dense atmosphere.

Colour Plate 7 is a photograph of Venus taken by Mariner 10 showing its cloud pattern, while Colour Plate 8 is a Venus close-up. It shows the *Ishtar terra* which is a high plateau of the size of Australia and contains several mountain ranges. The western part of Ishtar appears to be smooth plateau. It has been named the *Lakshmi Planum*.

The surface temperature is believed to be fairly high, in the range of 400 to 500°C, and almost constant over the whole planet. This leads to the possiblity of dust-laden storms and burning winds of fierce acid gases. Occasionally there would be pools of molten metal. All this excludes the possibility of any life.

However, it is because of this atmosphere that it reflects much of the light it receives from the sun.

SPACE PROBES

Since 1967 onward, several spacecraft have flown past Venus. Some Soviet space probes ejected canisters which have parachuted on its surface. The pictures taken from the space probes show that Venus is covered by mountains and valleys. The above data on temperature, pressure, atmosphere, mass and other parameters have been mostly obtained from photographs and data sent by these probes. (Also see Box 1.1.)

No appreciable magnetic field has been detected on Venus by these probes.

14

Mars, Jupiter and Saturn

In this chapter, we shall tell you about the three planets whose orbits lie outside that of the earth and which are visible to the naked eye. These are Mars, Jupiter and Saturn.

ASTRONOMICAL PARAMETERS

The mean distance of Mars from the sun is $a = 2.28 \times 10^8$ km (228 million km) or 1.524 AU. The eccentricity of its elliptical orbit is $e = 0.093$ which is the third highest among planets. Its farthest distance from the sun is 249 million km while its least distance from the sun is 207 million km. The aphelion is thus 1.205 times farther than the perihelion.

Colour Plates 9 and 10 show two views of Mars, one taken by Viking 1 from a distance of 56 000 km and the other, a close-up taken by Viking 2. Four volcanoes are visible on Martian surface in the former photograph.

Jupiter is the largest while Saturn is the second largest planet of the solar system. Jupiter is also usually the second brightest planet in the sky after Venus, although Mars at its maximum luminosity occasionally occupies the second place in this respect. Jupiter is brighter than Sirius which is the brightest star.

Colour Plate 11 shows Jupiter from a distance of 40 million km as observed by Voyager 1. Colour Plate 12 also shows another view of Jupiter.

Saturn appears yellowish to the naked eye. It was the outermost planet known till 1781, when Uranus was discovered.

Colour Plate 13 shows a marvelous view of Saturn from the other side of the solar system, a view that can never be obtained from the earth. It is a photograph taken by Voyager 1 four days after it flew past the planet. In this picture, the sun is to the right while the spacecraft is speeding to the left.

The semimajor axis (the mean distance from the sun) of Jupiter's orbit is 780 million km and that of Saturn is 1440 million km.

These and the other astronomical parameters of these three planets are given in Table 14.1.

The nearest distance of Mars from the earth is 56 million km. The maximum apparent size of this planet at such a favourable configuration is 25″. Such a favourable occasion occurs if at the time of opposition Mars is at perihelion while the earth is not far from its aphelion. It will be clear from Figure 12.1 that in such a case the distance SO and hence EO will be a minimum. Such an event occurs once in 15 to 17 years. The past two most

Table 14.1 The astronomical parameters of Mars, Jupiter and Saturn

	Parameter	*Mars*	*Jupiter*	*Saturn*
1.	Semimajor axis of orbit (mean distance from the sun), *a*, in million km	228	780	1440
	in AU	1.524	5.2	9.6
2.	Farthest distance from the sun (to aphelion, in million km)	249	817	1520
3.	Least distance from the sun (to perihelion, in million km)	207	742	1359
4.	Ratio aphelion distance/perihelion distance	1.203	1.101	1.118
5.	Eccentricity of orbit, *e*	0.093	0.048	0.056
6.	Orbital period, *P*, in years	1.88	11.86	29.458
7.	Inclination of orbit to ecliptic, *i* (in degrees)	1.85	1.3	2.5
8.	Average speed in orbit (in km/s)	24.1	13.2	9.66
9.	Apparent diameter (in seconds of arc), minimum	4″	32″	14″
	maximum	25″	50″	20″

favourable oppositions occurred in 1971 and 1986. Its minimum apparent size at conjunction is only 4″. Since Jupiter and Saturn are already far away from the earth, the variation in their distance at conjunction and opposition (or in their apparent maximum and minimum sizes) is not much.

All the three planets are exterior planets and hence show only gibbous phases. The maximum phase angle of Saturn is only 6° at quadrature, which means that Saturn almost always shows us its full face. The maximum phase angle of Jupiter and Mars at quadrature is 12° and 48°, respectively. (See Figure 12.3 for explanation.)

PHYSICAL PARAMETERS

Mars is visible as a bright red planet while Saturn appears yellowish in the night sky. Jupiter shows belts through a telescope. One can also see a large red spot on it through a telescope, known as the *Great Red Spot*. It has an oval shape and its size has been found to be 50 000 km along the long axis and 16 000 km along the short axis.

The Great Red Spot is surrounded by what scientists call a remarkably complex pattern of the giant planet's atmosphere. An elongated yellow cloud within the Great Red Spot can be seen swirling around the spot's interior boundary in a counterclockwise direction.

Saturn's rings are, of course, famous and they easily catch the attention of the viewer. Figure 14.1 shows the rings of Saturn at different inclinations. As saturn goes around the sun in its orbit, we sometimes see the rings edge-on, as it last happened in 1981 and will again happen in 1995. They are very difficult to observe at such times. The maximum inclination of the rings to the plane of the ecliptic is about 28°. The rings are most easily observed

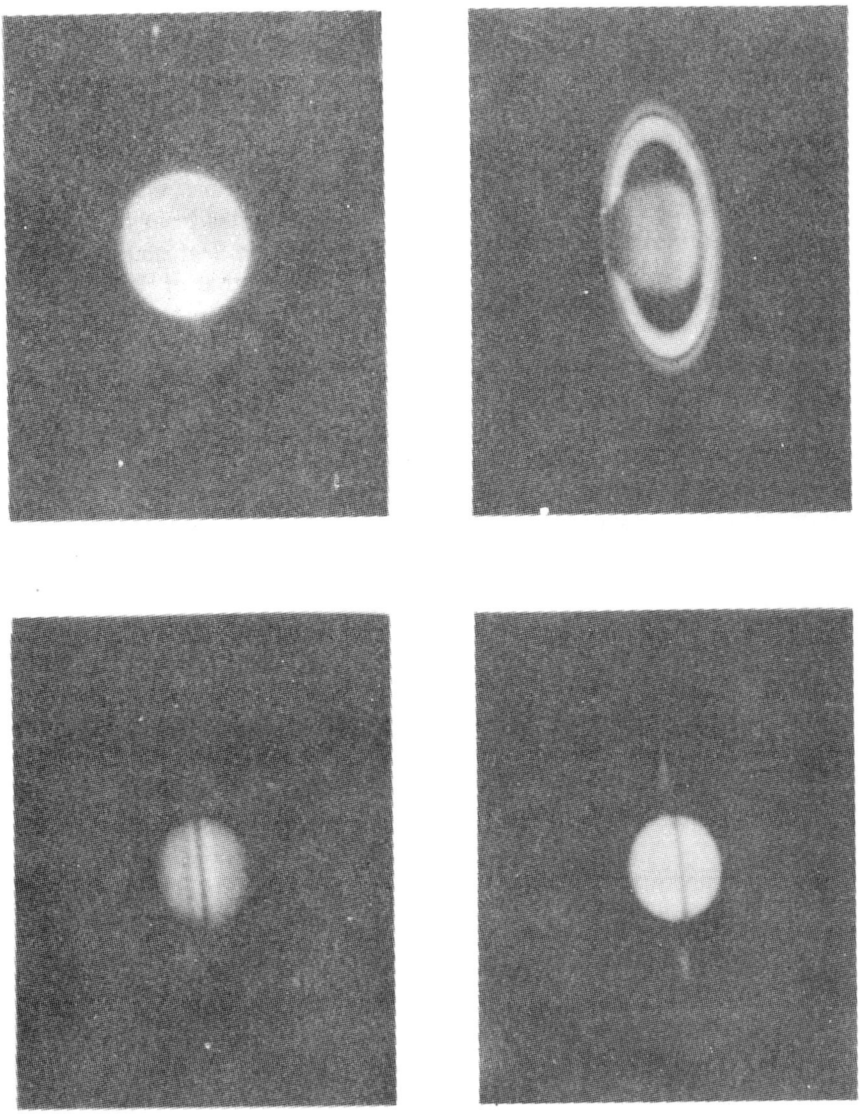

Figure 14.1 Rings of Saturn at different inclinations. The upper two photographs were exposed to show the planet (left) and the nearly edge-on rings (right). The inclination of the rings is the highest in the bottom right picture. (Photographed by 90 cm refractor; *Courtesy:* LO)

during such periods. The recent favourable period from this point of view was 1985-91.

The radius of Mars is 3398 km. The equatorial radius of Jupiter is 71 398 km while that of Saturn is 60 330 km. Their polar radii are considerably smaller than their equatorial radii.

The masses of these three planets have been determined fairly accurately from the motion of their satellites. The mass of Mars has been further calculated very accurately from the observations of Mariner 9 which orbited Mars in 1971-72. These are found to be about 10.75% of that of the earth for Mars, and 317.83 and 95.16 times that of the earth for Juptier and Saturn, respectively.

Mars rotates around itself in a period of 24 hours, 37 minutes and 22.1 seconds, a value which has also been measured very accurately. Jupiter and Saturn rotate around their axes at a much faster rate, respectively in 9 hr 55 min and 10 hr 40 min. You will thus see that the duration of a Martian day is very close to that on the earth. On the other hand, a Jovian day and a Saturnian day are much shorter, around 10 hours each. Jupiter, while being the largest planet of the solar system, also has the shortest period of rotation among all the planets.

Due to the fast rotation of Jupiter and Saturn about their own axes, these planets tend to bulge at the equator and flatten at their poles. The equatorial radius of Jupiter is 6.48% larger than its polar radius, while this figure is 10.76% in the case of Saturn.

The size (volume) of Jupiter is 1317 times that of the earth while that of Saturn is 762 times that of the earth. The mean density of Saturn is only about 70% of that of water, while the mean density of Jupiter is about the same as that of the sun.

The surface temperature on Mars is 50 °C lower than that on the earth. Due to the thin atmosphere which is unable to shield radiation from the sun, there is a large variation in temperature during the day and the night. On a winter night, the temperature is as low as – 100°C.

The surface temperature on Jupiter is about – 140°C and on Saturn around – 180°C. However, the temperature at the centre of Jupiter is believed to be around 30 000°C due to the presence of metallic hydrogen. Under severe pressure, electrons become free and the solid hydrogen there behaves like a metal. It is said that Jupiter just missed being a star. If it were a little more massive, it could have become a star. Jupiter contains a good deal of helium. Since this element is heavier than hydrogen, it is gradually settling down in the core and releasing heat. So it emits much more energy than it receives from the sun. Both Jupiter and Saturn fall into this category.

The axis of Jupiter's rotation around itself is inclined to the normal to its orbital plane only by 3°, while that of Saturn is inclined by 26° 45′. (Remember that in the case of the earth, this figure is 23° 27′.) Therefore Saturn shows a good deal of seasonal effects, like those on the earth. In the case of Jupiter, however, there is almost the same season throughout the Jovian year.

Table 14.2 lists information on the physical parameters of these planets.

Table 14.2 Physical parameters of Mars, Jupiter and Saturn.

	Parameter	*Mars*	*Jupiter*	*Saturn*
1.	Equatorial diameter (km)	6786.8	142 796	120 660
2.	Volume (in terms of earth volume)	0.150	1317	762
3.	Mass; in kg	6.4191×10^{23}	1.8988×10^{27}	5.685×10^{26}
	in terms of earth mass	0.1075	317.83	95.16
4.	Period of rotation (hr)	24.62296	9.92496	10.6567
5.	Inclination of rotation axis to normal to orbital plane	23°.98	3°.08	26°.73
6.	Albedo	0.15	0.43	0.61

ATMOSPHERE

Mars has an atmosphere made up chiefly of carbon dioxide and argon. A few other chemical species also exist in smaller proportions. These are carbon monoxide, oxygen, hydrogen and ozone. However, the Martian atmosphere is much thinner than our atmosphere. The atmospheric pressure on Mars in only about 1/20th of that in our atmosphere. There is clear evidence of clouds and dust-storms on the surface on Mars. Its albedo is 0.15, which means that it reflects only 15% of the light that falls on it.

Jupiter has an albedo of 0.43 while Saturn has 0.61. This is an indication of the fact that both these planets have a fairly thick atmosphere. Jupiter has an atmosphere about 1000 km thick, with a surface pressure three times the atmospheric pressure on the earth. The atmospheres of both Jupiter and Saturn contain methane and ammonia. It is dense and cloudy. In fact, Jupiter and Saturn, along with Uranus and Neptune, are called *gas giants* due to their vast thick atmosphere and lighter composition.

Box 14.1 *Spacecraft to Mars, Jupiter and Saturn*

The next two decades will see a large increase in space activity and space missions to study these three planets. The National Aeronautics and Space Administration of USA is planning to send a series of unmanned and manned spacecraft to Mars, beginning towards the end of the twentieth century. It is expected that man may land on this planet early in the twenty-first century.

NASA has sent a spacecraft to Jupiter in October 1989. It will arrive at Jupiter in December 1995. See Chapter 18 for more detail. NASA is also planning other missions to Saturn and its satellites beginning 1996. These would doubtless bring a wealth of information about these planets and satellites from close quarters. (See also Box 18.1.)

SATELLITES

Mars has two small satellites which orbit the parent planet very close to it. These two satellites, discovered in 1877 by Asaph Hall, are known as *Phobos* and *Deimos*. The radii of their orbits are 9350 km and 23 500 km, respectively. If you note that Mars itself has a radius of 3398 km, you will see that Phobos is only about 6000 km above its surface. The sizes of these satellites are extremely small (only a few tens of km) and they had not been determined very accurately until very recently. Moreover, both the satellites are far from spherical in shape. Their masses are also not known very accurately.

Colour Plate 14 shows the two Martian satellites, Phobos and Deimos, photographed by the Viking spacecraft. Note the haphazard shape of both the satellites, like broken bodies. This must have been due to the impact of other celestial bodies which roam around in interplanetary space.

Man may very well land on these satellites during the next two decades. Our astronaut will have to anchor himself fast to either of these satellites. Their gravity is so small that even a small kick on the surface will make our astronaut fly into space.

There is another interesting thing about Phobos to be noted. Its period of revolution about Mars (7 hr 39 min) is less than one-third of the period of rotation of Mars about its own axis (24 hr 37 min). Phobos runs faster in its path around Mars than a point on the Martian equator. It would therefore appear to a creature on Mars to rise in the west and set in the east. It is the only known natural satellite in the solar system to have a period of revolution smaller than the period of rotation of the parent planet.

As of today, Jupiter has 16 satellites while Saturn has 18. The larger four satellites of Jupiter, known as Io, Europa, Ganymede and Callisto, were discovered by Galileo in the year 1610. The largest satellite of Saturn, Titan, was discovered in 1655, while its tenth satellite was discovered in 1966. Its thirteenth satellite was discovered in 1974 and the others have been discovered later. Titan, which has a diameter of about 5000 km, is larger than the planet Mercury and is only a little smaller than Mars. It is the *only satellite* in the solar system to have an atmosphere. Some information on the satellites of these planets is collected in Table 14.3.

Colour Plate 15 is a collage of photographs of Jupiter and its four Galilean satellites, Io, Europa, Ganymede and Callisto, taken by Voyager 1.

Colour Plate 16 is a montage of photographs of Saturn and six of its satellites as observed by Voyager 1.

The newer and smaller satellites have been discovered by two spacecraft. Voyager 1 and Voyager 2 flew past Jupiter in March 1979 and July 1979, respectively. They flew past Saturn in November 1980 and August 1981, respectively. (See Chapter 18.)

Table 14.3 Satellites of Mars, Jupiter and Saturn*. Some satellites are of very irregular shape. The linear dimensions are mentioned in their case

Satellite	*Semimajor axis (in 1000 km)*	*Period of revolution around planet (in earth days)*	*Radius/size of satellite (in km)*
(a) *Mars*			
1. Phobos	9.378	0.319	13.5×10.8×9.4
2. Deimos	23.459	1.26	7.5×6.1×5.5
(b) *Jupiter*			
1. Metis	128	0.295	20
2. Adrastea	129	0.298	12.5×10×7.5
3. Amalthea	181	0.498	135×83×75
4. Thebe	222	0.675	~ 50
5. Io	422	1.77	1815
6. Europa	671	3.55	1569
7. Ganymede	1070	7.15	2631
8. Callisto	1883	16.7	2400
9. Leda	11 094	239	8
10. Himalia	11 480	251	93
11. Lysithea	11 720	259	18
12. Elara	11 737	260	38
13. Ananke	21 200	631 (R)#	15
14. Carme	22 600	692 (R)	20
15. Pasiphae	23 500	735 (R)	25
16. Sinope	23 700	758 (R)	18
(c) *Saturn*			
1. Atlas	137.7	0.602	20×10
2. Prometheus	139.4	0.613	70×50×40
3. Pandora	141.7	0.629	55×45×35
4. Epimetheus	151.4	0.694	70×60×50
5. Janus	151.5	0.695	110×100×80
6. Mimas	186	0.942	196
7. Enceladus	238	1.37	250
8. Tethys	295	1.89	530
9. Telesto	295	1.89	14
10. Calypso	295	1.89	17×11×11
11. Dione	377	2.74	560
12. Helene	377	2.74	18×16×15
13. Rhea	527	4.52	765
14. Titan	1222	15.9	2575
15. Hyperion	1481	21.3	205×130×110
16. Iapetus	3561	79.3	730
17. Phoebe	12 952	550 (R)	110
18. Pan	?	?	10

*Data based on *The Astronomical Almanac of 1991*, published by the International Astronomical Union.
~ indicates that the value is approximate.
# (R) indicates retrograde motion of the satellite, opposite to the daily motion of the planet.

15

The Asteroids

We have now studied six planets of the solar system nearest to the sun—the earth and five other planets. We have mentioned earlier that these five planets were the only known (major) occupants of the solar system, apart from the earth and the sun, until the year 1781, when the next planet Uranus was discovered. We will come to this and the other outer planets of the solar system in the next chapter.

We must discuss in this chapter other interesting developments leading to the discovery of the asteroids. The history of these discoveries makes fascinating reading.

Figure 15.1 shows the orbits of five asteroids, Ceres, Apollo, Eros, Hermes and Hidalgo. Note that the orbit of Ceres is almost circular, like that of a planet.

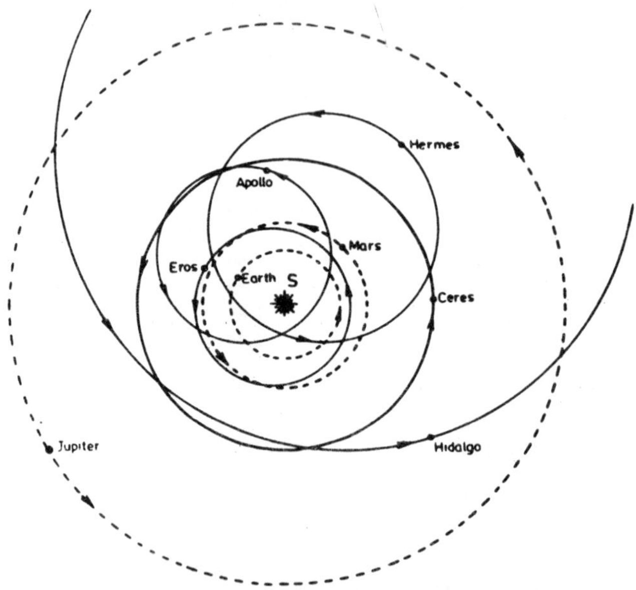

Figure 15.1 Orbits of a few asteroids. Also shown are the orbits of the earth, Mars and Jupiter.

BODE'S LAW

Some astronomers of the eighteenth century with a mathematical bend of mind had tried to find relationship in the mean distances of these planets

Box 15.1 *Indian scientist discovers new asteroid*

An Indian astronomer, Dr. R. Rajamohan, of the Indian Institute of Astrophysics, Bangalore, has discovered a new asteroid. The announcement was made in news media in December 1989. Other members of the discovery team include JC Bhattacharyya, K Kuppuswamy, A Paranjpye, V Moorthy, and R Vasundhara.

The new asteroid was first spotted on February 17, 1988, in photographs taken by the 45 cm telescope of the Vainu Bappu Observatory at Kavalur (see Figure 15.2). It was under observation for more than a year before its discovery was announced.

It is the four thousand one hundred and thirtieth asteroid to be catalogued. It has been named after the great Indian mathematician Srinivasa Ramanujan, according to the wishes of the discoverer. Thus its catalogue name is (4130) *Ramanujan* asteroid.

When the new asteroid was discovered by R Rajamohan, it was given the temporary name 1988 DQ1. The measurements of its position were carried out by A Paranjpye. The object was followed up and totally five positions were reported to the Minor Planet Centre (MPC). A preliminary orbit for this object was reported by S Nakano in May 1988. R Vasundhara developed programmes for predicting its position on future dates based on the provisional orbit given by Nakano. 1988 DQ1 was recovered close to its predicted position on plates taken at Kavalur on May 4, 5, 6 and 7, 1989. These observations led Nakano to derive firm orbital elements and the MPC assigned number 4130 to this object in July 1989.

from the sun. Among these, the relation obtained by Titius and Bode is famous. On the basis of strange logic, these German astronomers obtained a formula for the distances of planets from the sun. Bode's relation is given in Table 15.1. (See also Box 15.2.)

Table 15.1 Bode's relation for distances of planets from the sun and their actual distances.

Planet	*Distance from the sun (in AU)*	
	According to Bode's law	*Actual (average)*
1. Mercury	0.4	0.39
2. Venus	0.7	0.72
3. Earth	1.0	1
4. Mars	1.6	1.52
5. Missing planet	2.8	(asteroids)
6. Jupiter	5.2	5.20
7. Saturn	10.0	9.55
Planets discovered after enunciation of Bode's law		
8. Uranus	19.6	19.22
9. Neptune	38.8	30.11
10. Pluto	77.2	39.46

When the seventh planet of the solar system, Uranus, was discovered in 1781, several astronomers rightly thought that there could be many other planets in the solar system, not seen until then.

Box 15.2 *The logic of Bode's law*

Take the series 0, 3, 6, 12, 24, 48, 96, ..., starting with 0 and 3 and writing each successive number by doubling the previous one. Add 4 to each term to get the series 4, 7, 10, 16, 28, 52, 100, ... Now divide each number by 10 to obtain the new series 0.4, 0.7, 1.0, 1.6, 2.8, 5.2, 10.0, ... According to Bode, these are the distances of the planets from the sun in terms of the earth-sun distance (1 AU).

We shall give here the mathematical equation which describes Bode's law, according to the above procedure. The semimajor axis of the *n*th planet, in astronomical units (AU), is given by

$$a\ (\text{AU}) = \begin{cases} 0.4 \text{ for } n = 1 \\ 0.4 + 0.3 \times 2^{n-2} \text{ for } n = 2, 3, 4, \ldots \end{cases}$$

When Uranus was discovered in 1781 at a mean distance of 19.22 AU from the sun, a new term was added to the above series. According to the above rule, you can see that the next term comes out to be 19.6, which was close to the actual value. This agreement led astronomers to think that there must be some deep significance to the Titius-Bode law.

Although the actual distances of planets up to Uranus agree very well with those given by Bode's law, one discrepancy is obvious—there is a gap between Mars and Jupiter. *Has the divine creator,* argued the astronomers of the late eighteenth century, *left this gap empty?* An emphatic *no* was their reply.

IT STARTED WITH CERES

In spite of the fact that Bode's law was not built upon scientific logic or reasoning, it did serve the important purpose of focussing the attention of the astronomers of the day to this region of the cosmos. On the new year day of 1801, G. Piazzi, in the Sicily Islands of Italy, noticed a tiny spot of light in the sky. He could not identify it with any known star. The next day, the spot of light was not there, but there was a similar *star* nearby which was not there on the previous day. On January 3, 1801, this spot had further shifted its position. It thus could not be a star.

Piazzi thought that it was a comet. He informed Bode in Germany about his discovery while continuing to keep his eye on the new object night after night. Bode made a few calculations and declared that the search for a planet between Mars and Jupiter was successful. According to the established practice in astronomy, it was Piazzi's privilege to name the new object. He chose the name *Ceres* (from Greek mythology) for the first asteroid.

Figure 15.2 shows two photographs by which the (4130) Ramanujan asteroid was discovered. The asteroid is marked by arrows in both the

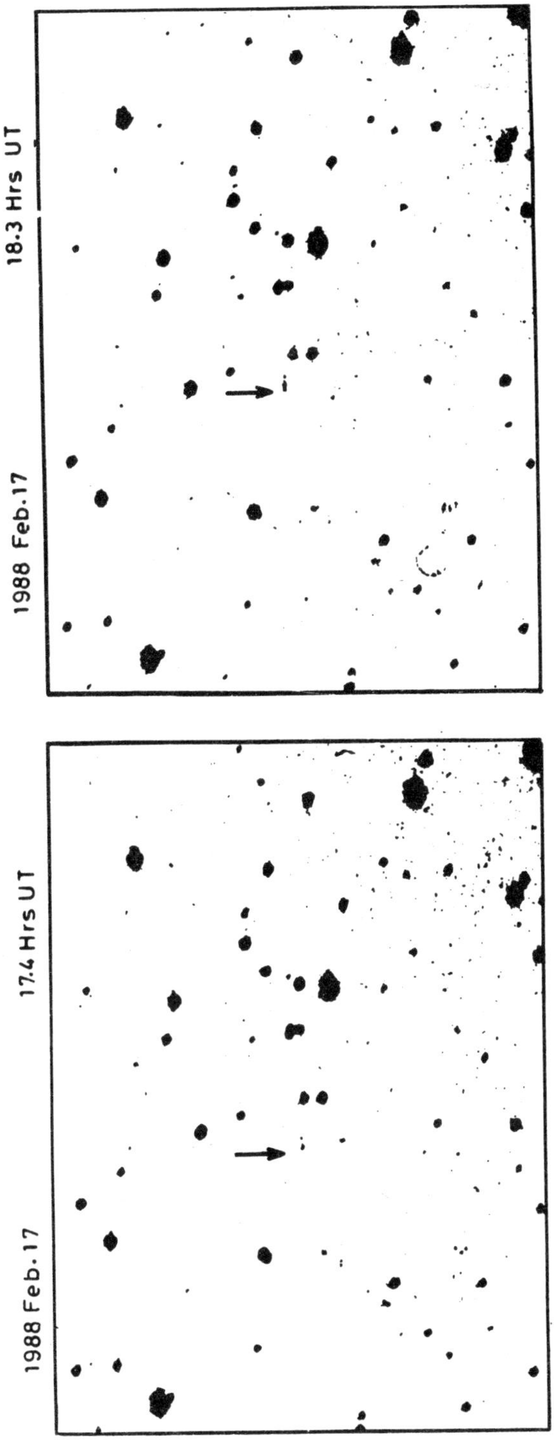

Figure 15.2 Two photographs heralding the discovery of the new (4130) Ramanujan asteroid. Note the slight shift in the position of the object marked by the arrow against the starry background (*Courtesy;* IIA).

photographs. You will notice that the object shown by the arrow has shifted its position slightly with respect to the stars in a span of less than an hour. (See also Box 15.1.) This is similar to Figure 2.1 showing the motion of Pluto. In another method, an asteriod leaves a streak of light on a time-exposure photograph as against the sharp images of stars.

MATHEMATICAL ADVANCES

Ceres was lost after a few nights of observation due to bad weather. It was as if nothing was known about this minor planet except its name.

A few years before, towards the end of the eighteenth century, tremendous mathematical advances had been made in the field of astronomy. Gauss, a great contemporary mathematician, found out a way to calculate the orbit of a celestial body *with only a few* observations, on the basis of Newton's laws of gravitation and Kepler's laws of planetary motion. He applied his method to Ceres, and calculated the orbit it must follow by using the few positions which were known from observations in January 1801. Two astronomers, von Zach and W Olbers, rediscovered the asteroid in December 1801 at a position predicted by the orbit calculated by Gauss.

It was a triumph for the ever-growing relationship of mathematics, physics and astronomy.

MORE ASTEROIDS

On March 28, 1802, W Olbers discovered the second asteroid, named *Pallas* at about the same distance from the sun. This was unbelievable. Were there two planets between Mars and Jupiter, orbiting the sun?

The wonder was only short-lived. On 1st September, 1804, K Harding discovered the third asteroid, called *Juno*. On March 29, 1807, Olbers again discovered the fourth asteroid, called *Vesta*. This was not an accident but the discovery was a result of systematic search of the sky by Olbers.

People had to wait for more than 38 years for the discovery of the fifth minor planet. This planet was discovered on December 8, 1845, and was named *Astrea*.

It soon became almost an astronomical sport to search for new asteroids. Hardly a year has passed since then without the discovery of a new asteroid. The hundredth asteroid was discovered in 1868, and about 500 asteroids were known by the turn of the century, that is by the year 1900.

The discovery of these asteroids makes an interesting reading. A German astronomer Goldschmidt, living in Paris, discovered 14 asteroids, some of them from the window of his living room. Luther discovered 24 asteroids, while J Palisa has more than 60 asteroids to his credit.

More than 4000 asteroids are now known. In fact, the number stands at 4865 at the end of July 1991.

NEW METHODS OF OBSERVATION

In the nineteenth century, the observation of the sky was mainly done through the telescope. But very soon newer methods of observation were coming on the scene.

In 1891, Max Wolf discovered an asteroid by the photographic method for the first time. Since then there has been a considerable improvement in the method of observation and new techniques have arrived. We now have telescopes of much higher resolution and very sensitive photographic methods.

DISCOVERY OF ASTEROIDS AND NOMENCLATURE

With the improvement in mathematical and computational techniques, scientists can determine the exact orbit of a body in the solar system from *only three* observations. When an asteroid is observed at least on three occasions, it is given a serial number and a name. For example, (1002) *Olbersia* indicates that it is the one-thousand-and-second asteroid to be discovered and is known by the name Olbersia. The one-thousandth asteriod to be discovered has been named after Piazzi, the discoverer of the first asteroid. Hence it is denoted by (1000) *Piazzia*. The 5000th asteroid is now proposed to be named *IAU*, after the International Astronomical Union.

When a new asteroid is sighted for the first time, it is given a provisional name, such as 1957 NA. Only when it is sighted at least three times is it given a permanent name and entered into the catalogue. See also Box 15.1.

ASTRONOMICAL AND PHYSICAL PARAMETERS

Asteroids are very small bodies when compared to the planets in the solar system. The largest asteroid, Ceres, which was also the first to be discovered, has a diameter of only about 1003 km. The sizes of the largest four asteroids are given in Table 15.2. Their sizes, as compared to the moon (the natural satellite of the earth), can be seen in Figure 15.3. If the diameter of the moon is taken as 100 units, the diameters of the four asteroids would be: Ceres: 28.8 units; Pallas: 17.5 units; Vesta: 15.5 units; Juno: 7.1 units.

Table 15.2 The diameters of the largest four asteroids

Asteroid	*Serial number of discovery*	*Diameter (km)*
1. Ceres	1	1003
2. Pallas	2	608
3. Vesta	4	538
4. Juno	3	247

The orbital elements of the four major asteroids are given in Table 15.3. Note that in this table, the last entry is the average angle n traversed by the asteroid in the sky per day. This can be easily related to the period of revolution P in days. Noting that an object covers 360° (a complete orbit) during the period P, we would have

$$nP = 360° \quad \text{or } P = 360°/n \quad \text{or } n = 360°/P.$$

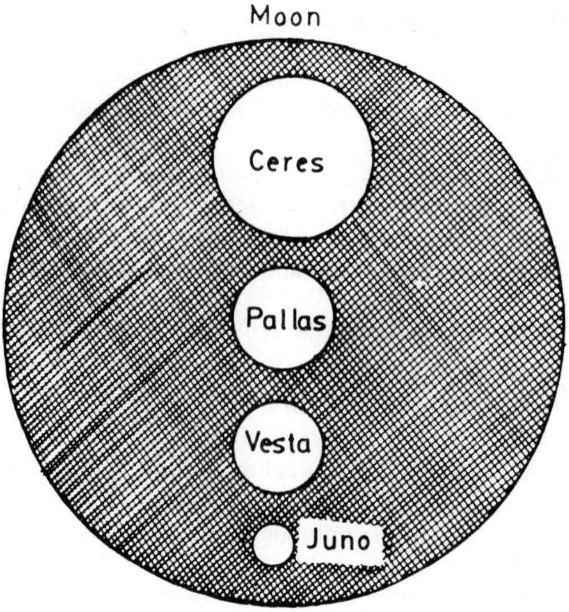

Figure 15.3 The largest four asteroids, Ceres, Pallas, Vesta and Juno, superimposed on the moon to show their relative sizes.

Table 15.3 The orbital elements of the four major asteroids (reference: ecliptic and equinox of 2000). The mean angular velocity is with respect to the sun (heliocentric).

	Orbital element	*Ceres*	*Pallas*	*Juno*	*Vesta*
1.	Argument of perihelion, ω (degrees)	71.204	309.796	246.744	150.175
2.	Longitude of ascending node, Ω (degrees)	80.702	173.323	170.556	104.015
3.	Inclination of the orbit, i (degrees)	10.607	34.804	12.993	7.139
4.	Eccentricity of orbit, e	0.0780	0.2347	0.2580	0.0906
5.	Semimajor axis, a (AU)	2.7685	2.7703	2.6682	2.3613
6.	Period of revolution around the sun, P (years)	4.6065	4.6110	4.3584	3.6285
7.	Mean angular velocity per day, n (degrees)	0.21396	0.21375	0.22614	0.27163

For example, Ceres has a peiod of 4.61 years, or 1682 days. Therefore its mean daily motion would be

$$n = 360°/1682 \text{ days} = 0°.214 \text{ per day.}$$

We give here the orbital elements and the apparent magnitude of the new (4130) Ramanujan asteroid:

a = 3.053 940 5 AU, e = 0.044 536 4, i = 9°.790 49,
Ω (reference 1950) = 53°.200 11, ω = 174°.983 29,
P = 5.34 years, T = October 1, 1989,
Apparent magnitude at 1988 opposition = 14.3.

In addition to the four major asteroids listed in Tables 15.2 and 15.3, there are numerous other asteroids, as we have mentioned earlier. There seems to be a lower limit to the length of the semimajor axis of the asteroids. There are a few asteroids for which the length of the semimajor axis is less than 1 AU, or orbital period less than one year. No upper limit to this parameter has yet been found with any certainty. Thus asteroids have been discovered whose orbits extend beyond the orbit of Saturn.

WHAT HAVE THE ASTEROIDS GIVEN US?

Asteroids are extermely small bodies as compared to the major planets and satellites of the solar system. Why is it important to study them?

Physicists and astronomers are faced with several problems. They must explain how the solar system was formed, did the planets evolve from the primordial gas or were they captured by the sun, and similar questions about comets, their physical properties, the inclination of their orbits, etc. Because they are different from the major planets and satellites, at least in their sizes, the asteriods can be used to test the theories propounded by the scientists.

An altogether different kind of gain has been the accurate measurement of the earth-sun distance (the astronomical unit, 1 AU) with the help of an asteroid. The asteroid Eros came close to the earth twice during the twentieth century, first during 1900-01 and then during 1930-31. On these occasions, it came to within 22 million km of the earth. (Note that 1 AU is about 150 million km.)

Careful observations of the motion of Eros lead to a very accurate determination of this astronomical unit of distance.

The Galileo spacecraft, which has been recently launched towards Jupiter, will pass by an asteroid on its way. (See Box 18.1.)

Box 15.3 *Bringing an asteroid to the earth*

Some scientists have made a fantastic and far-reaching suggestion which, at this moment, may appear to be straight from science fiction. According to this suggestion, a small asteroid may be brought near the earth and made to orbit it as its satellite.

The suggestion is to land on a small asteroid and fire jets to change its speed so that it moves towards a lower orbit. Once near the earth, engines would again be fired to put it in an orbit around the earth.

It could very well be within the powers of man to act on this suggestion in 50 years or so.

If this is done, it would be possible to study its contents more closely. This may throw light on the evolution of the solar system and the formation of asteroids.

16

The Outermost Planets—Uranus, Neptune and Pluto

Now we come to the three outermost planets of the solar system. Figure 16.1 shows Uranus with some of its satellites while Figure 16.2 shows Neptune with its satellites.

DISCOVERY OF THE THREE PLANETS

The discovery of the three planets, Uranus, Neptune and Pluto, makes an interesting reading.

Until 1781, nobody suspected that the sun had a planet beyond the orbit of Saturn. But on March 13, 1781, W Herschel disovered a distant planet beyond Saturn. Subsequently, this was called *Uranus*. It was the first planet to be discovered with a telescope.

Astronomers had earlier observed some irregularities in Saturn's motion. On certain occasions, it was speeding up and then slowing down, as compared to its expected speed. After the discovery of Uranus, these irregularities were explained as being due to the gravitational pull of Uranus whenever Saturn happened to be on the same side as Uranus.

Once Uranus was discovered, its orbit was calculated and its motion carefully studied. It turned out that its motion was also irregular, during the first few decades of the nineteenth century. Astronomers rightly thought that there must be a planet beyond Uranus which is causing these perturbations.

This time, calculations were made around 1845-46, independently by John Adams in the UK and Urbaine Le Verrier in France, from the discrepancies in the motion of Uranus. The mass, the orbital elements, and the expected position of the unknown planet were predicted. Soon after, *Neptune* was discovered on September 23, 1846, by JG Galle within a degree of the predicted position. This was a great achievement for celestial mechanics, the branch of mathematical physics dealing with the motion of heavenly objects.

The orbit calculations of Neptune showed that Uranus and Neptune were in opposition around the year 1822. It was this period, a little before and after 1822, when the motion of Uranus showed irregularities which led to the prediction of Neptune. It is interesting to note that their opposition occurs after *several hundred years*. Thus if the discrepancies in the motion of Uranus had not been noticed by astronomers in the first few decades of the

Figure 16.1 Uranus and some of its satellites. (*Courtesy:* LO)

nineteenth century, there is good reason to suppose that the discovery of Neptune would have been considerably delayed.

PLUTO'S DISCOVERY AND THE ANTICLIMAX

The story repeats itself in the case of Pluto, though with an important difference. Discrepancies were observed in the motion of Neptune in the early part of the twentieth century. This led to the expectation that there was a planet beyond Neptune. The mass, the orbital elements, and the expected position of the new planet were predicted in the beginning of this century by an Indian, VB Ketkar, and later in 1915 by Perceival Lowell.

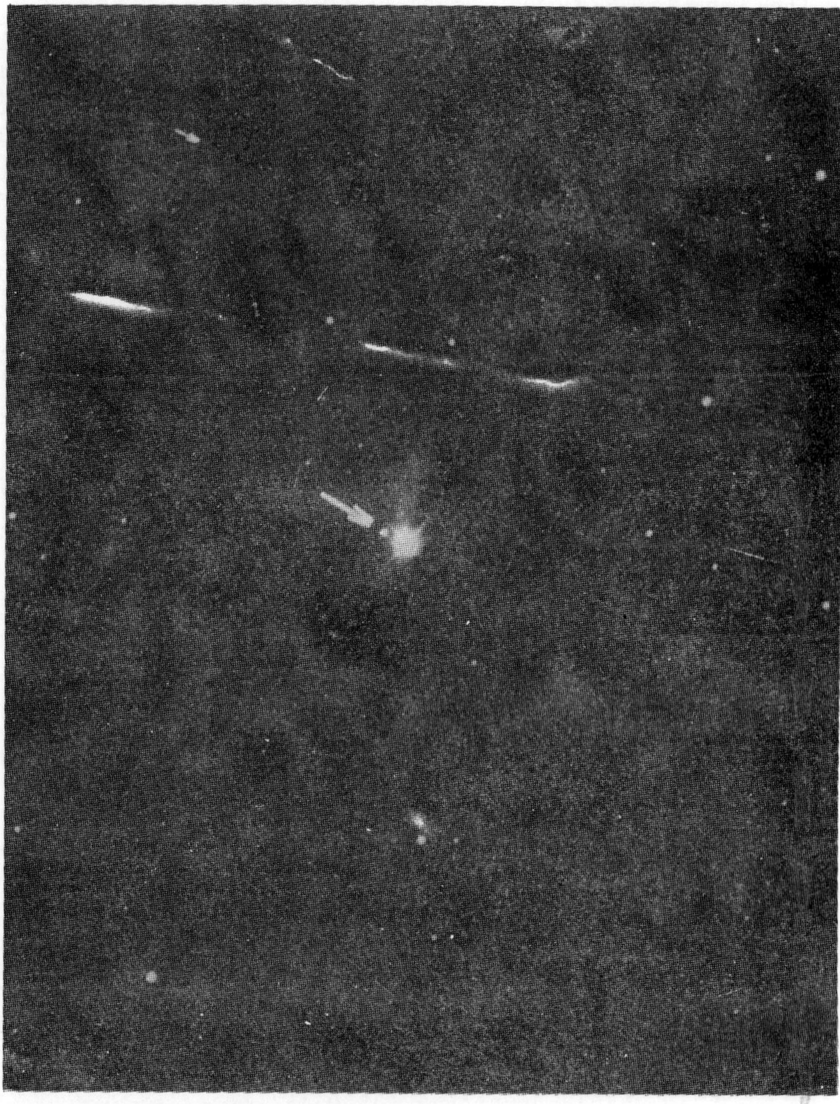

Figure 16.2 Neptune and some of its satellites. (*Courtesy:* LO)

The distance of this ninth planet from the sun was so large that it could not be observed by the telescopes available then. A new wide-field telescope was constructed particularly for the search of a trans-Neptunian planet. This was completed and put in operation in 1929. Several photographs of millions of stellar objects were taken by this telescope in the region of the sky where the new planet was expected to be. Finally, CW Tombaugh discovered *Pluto* on February 18, 1930, by comparison of two photographic plates taken at a few days' interval.

Then came the anticlimax, though 50 years later, when its mass was estimated directly by the discovery and observation of its natural satellite. Although the orbital elements of the new planet matched with those predicted, its mass fell short by a factor of 100. The mass and size of this new planet turned out to be so small (smaller than that of the earth) and its distance from Neptune, even at opposition, so large (about the distance between the earth and Uranus), that it was impossible to believe that this was the planet disturbing Neptune in its path. We may therefore say that the discovery of Pluto was just a happy accident.

We have earlier shown, in Figure 2.1, two photographs of Pluto showing its motion in a day against the starry background.

Some astronomers believe that there is still another, a tenth, planet not yet discovered. They have given it the name Planet X and the search continues. In fact, a project named *Kalki* Project has been launched by astronomers at Bangalore and Kavalur in January 1987 to search for the tenth planet. (*Kalki* is the name of the tenth incarnation in Indian mythology.) It was during this search that they discovered a new asteroid. (See Box 15.1.)

ORBITAL ELEMENTS

The orbital elements of these three planets are given in Table 16.1. You can notice that Pluto is different from the other two planets. Its orbit is highly eccentric, so much so that its smallest distance from the sun is only 29.81 AU while is farthest distance from it is 49.26 AU, 1.66 times the former. (See also Table 11.1.)

Table 16.1 The orbital elements of Uranus, Neptune and Pluto

	Orbital element	*Uranus*	*Neptune*	*Pluto*
1.	Semimajor axis,			
	a, in AU	19.183	30.055	39.44
	in 10^9 km	2.870	4.497	5.900
2.	Eccentricity of orbit, e	0.047	0.0086	0.247
3.	Period of revolution, P (years)	84.02	164.77	247.7
4.	Inclination of orbit, i (degrees)	0.772	1.771	17.15
5.	Mean speed in orbit (km/s)	6.81	5.45	4.73

On the other hand, the mean distance of Neptune from the sun is 30.1 AU with an almost circular orbit (very small eccentricity). This means that at its perihelion, Pluto is nearer to the sun than is Neptune. This is, in fact, the situation at the moment of writing this (1990). Their orbits, however, do not intersect due to the large angle between the orbits of the two planets. Fortunately, therefore there is no chance of planetary collision.

A sketch of the orbits of the three planets is shown in Figure 16.3. Although Pluto's orbital plane is inclined to the plane of the ecliptic by 17°, its projection has been shown. The thin lines connecting two orbits indicate that the concerned planets were or would be at those positions in their respective orbits in the years shown.

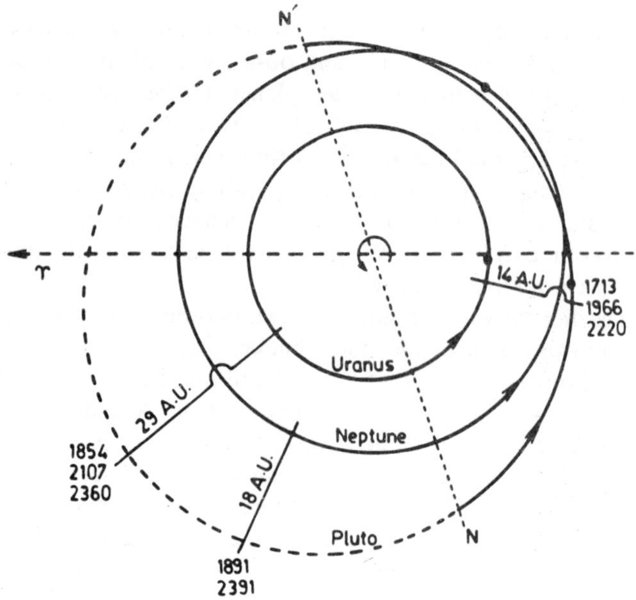

Figure 16.3 Projection of Pluto's orbit in relation to those of Uranus and Neptune. N and N′ are the ascending and descending nodes of Pluto, respectively, and ♈ indicates the direction of the vernal equinox. Dots indicate the positions of the three planets in 1967. (*Courtesy*: EB. Reproduced with permission from "Pluto" in *Encyclopaedia Britannica,* 15th edition, 1974.)

You may also note that twice the period of revolution of Pluto around the sun is approximately equal to thrice the period of Neptune, and both of these are very close to 500 years. Therefore, once in every 500 years, Pluto and Neptune can come very close to each other in their orbits. On such occasions, the distance between them is only 18 AU. Although this is much smaller than their mean distance from the sun, it is larger than the distance of Saturn from the earth. The next such event will occur in the year 2391.

Similarly at close conjunction, Uranus and Pluto are only 14 AU apart. This occurred most recently in 1966 and will occur again in the year 2220.

Pluto's orbit is highly inclined to the ecliptic. The angle of inclination is $i = 17°.15$. This has given rise to various speculations among scientists about the origin of this planet.

PHYSICAL PARAMETERS

Uranus has several satellites while Neptune also has a few. Therefore, the masses of these two planets have been known with fairly good accuracy. Pluto's mass was very difficult to calculate until very recently. However, one satellite of Pluto has been discovered recently and we have some idea of its mass. There are still many uncertainties in this regard for the reason that it is extremely difficult to observe the satellite of this small and distant planet.

The physical parameters of these planets are listed in Table 16.2. Uranus and Neptune are large planets, like Jupiter and Saturn, while Pluto is very small, much smaller than Mercury. An interesting feature is the high angle, about 97°.9, of the axis of rotation of Uranus to its orbital plane. Thus Uranus appears as if it is rolling on its sides while going around the sun, as one rolls on the grass.

Uranus and Neptune have a dense atmosphere of poisonous gases like hydrogen, methane and ammonia. The surface temperature is estimated to be about 60 K (– 214°C) for Uranus, about 53 K for Neptune, and 68 K for Pluto.

Table 16.2 Physical parameters of Uranus, Neptune and Pluto

Parameter	*Uranus*	*Neptune*	*Pluto*
1. Diameter (km)	51 118	49 500	2245
2. Mass (units of earth masses)	14.501	17.239	0.00228
3. Period of rotation about own axis (hours)	17.24 (R)	16.05	153.29 (R)
4. Inclination of the axis of rotation (degrees)	97.86	29.56	118
5. Albedo	0.35	0.41	0.5

(R indicates retrograde motion)

Voyager II has discovered a peculiar spot on Neptune. Figure 16.4 is a photograph of Neptune taken by Voyager II, showing the spot on its surface.

Colour Plate 17 is a painting of Pluto and its satellite Charon. Even in the world's largest telescope, Pluto is seen as a tiny dot. It has a diameter of 2245 km, its satellite Charon has a diameter of 1120 km, and the separation between them is 19 130 km.

ATMOSPHERE

Uranus and Neptune are believed to have a huge thick atmosphere of poisonous gases. In this respect, they are similar to Jupiter and Saturn. There is a good deal of similarity in their sizes too. These four planets, Jupiter, Saturn, Uranus and Neptune, are the largest four planets in the solar system. They are therefore called the *gas giants*.

RINGS OF URANUS

If you look through a simple telescope, you can see the rings of Saturn alone. However, with better techniques and spacecraft observations, Jupiter, Uranus and Neptune have also been found to possess rings around them.

Figure 16.5 is a sketch of Uranus and its rings. The markings on the diagram indicate where or by which group the rings were discovered. Two observatories in India, at Naini Tal and Kavalur, have contributed to this search using their telescopes. Like Saturn's rings, those of Uranus are believed to consist of a swarm of small particles.

Colour Plate 18 is an artist's conception of Uranus and its rings. It shows nine known rings, extending from 15 000 to 26 000 km in width.

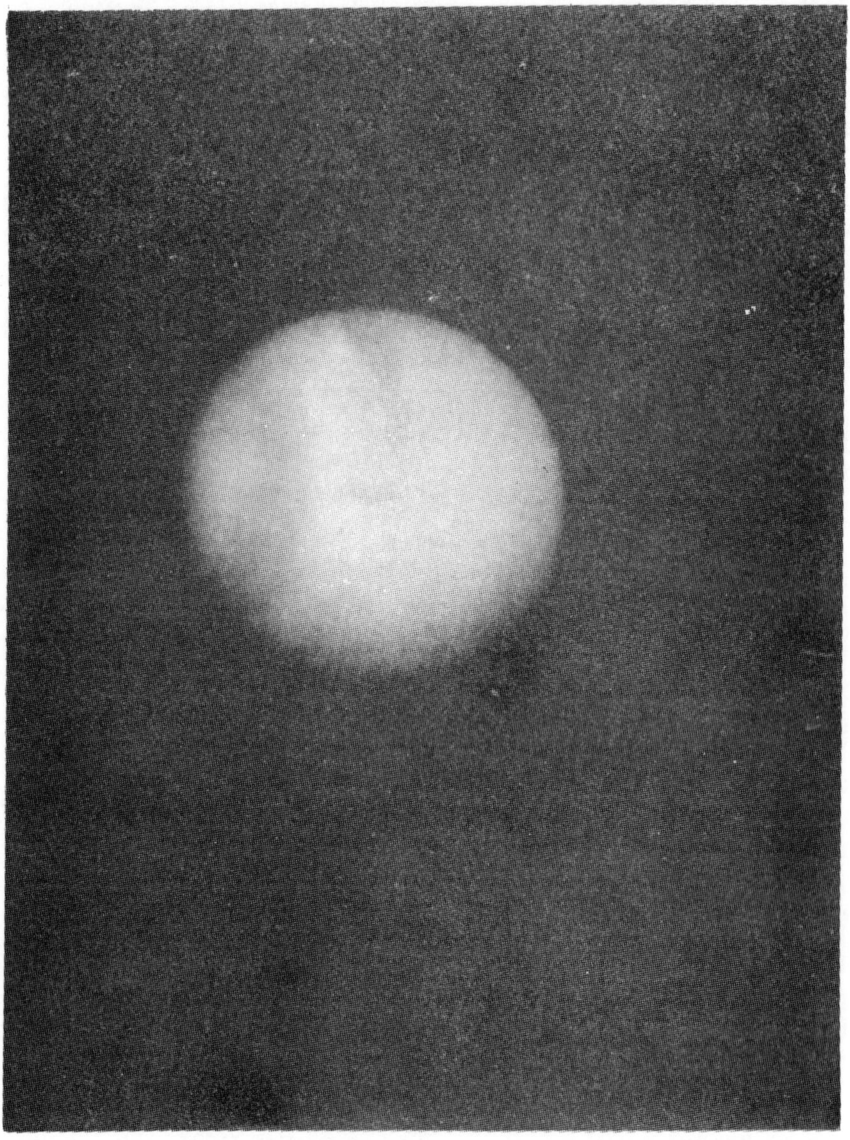

Figure 16.4 Neptune with a spot on its surface. (*Courtesy*: TG)

SATELLITES

The spacecraft Voyager II flew past Uranus in 1986 and past Neptune only in August 1989. It was nearest Neptune on August 24, 1989. It is the first man-made spacecraft to go near these distant planets.

It has observed several new moons of these two planets. The numbers of their satellites now stand at 15 and 8, respectively.

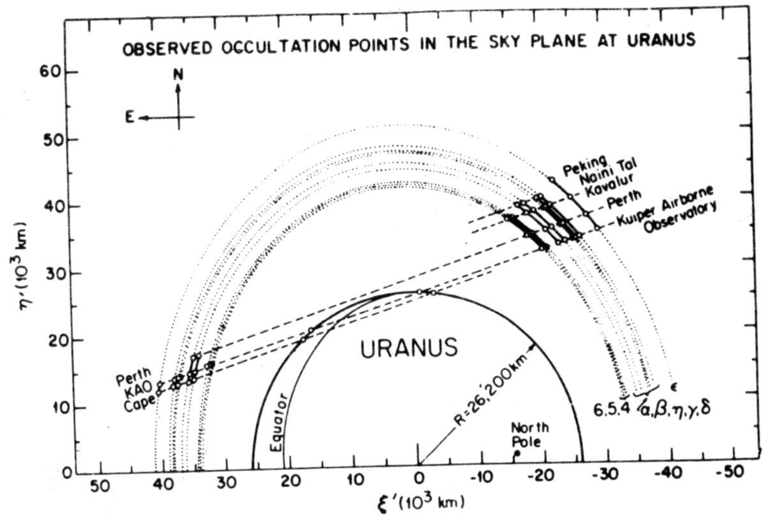

Figure 16.5 Uranus and its rings; the various laboratories which discovered the rings are mentioned. (From *The Astronomical Journal 83, 980 (1978);* with the permission of authors JLE, LHW, RLM and JC.)

Colour Plate 19 is an artist's conception of the planet Neptune as seen from its largest satellite Triton.

The only known satellite of Pluto, *Charon,* was discovered in 1978. It seems that the planet Pluto is only twice its satellite in linear size. The pair of Pluto and Charon is almost like a double-planet, revolving around their common centre of mass.

Table 16.3 gives several features of these satellites.

Table 16.3 Satellites of Uranus, Neptune and Pluto*

Satellite	*Semimajor axis (in 1000 km)*	*Period of revolution around planet (in earth days)*	*Radius of satellite (in km)*
(a) *Uranus*			
1. Belinda	49.77	0.335 03	25
2. Cordelia	53.79	0.376 41	25
3. Ophelia	59.17	0.434 58	25
4. Desdemona	61.78	0.463 57	30
5. Rosalind	62.68	0.473 65	30
6. Cressida	64.35	0.493 07	40
7. Bianca	66.09	0.513 20	40
8. Juliet	69.94	0.558 46	30
9. Portia	75.26	0.623 53	25
10. Puck	86.01	0.761 83	85
11. Miranda	129.39	1.413 48	240
12. Ariel	191.02	2.520 38	579
13. Umbriel	266.30	4.144 18	586

(Contd.)

Table 16.3 contd.

14. Titania	435.91	8.705 87	790
15. Oberon	583.52	13.463 24	762
Neptune			
1. 1989 N6	48.2	0.29	25
2. 1989 N5	50.0	0.31	45
3. 1989 N3	52.0	0.34	70
4. 1989 N4	62.0	0.43	80
5. 1989 N2	73.0	0.56	100
6. 1989 N1	117.5	1.1223	210
7. Triton	354.29	5.876 84 (R)#	1900
8. Nereid	5511	360.2	150
Pluto			
1. Charon	19.13	6.387 18	559.7

*Data based on *The Astronomical Almanac of 1991* published jointly by the US Naval Observatory and Royal Greenwich Observatory, and the IAU Circular Nos. 4806, 4824 and 4867 issued by the International Astronomical Union.

Retrograde motion opposite the daily rotation of the planet.

17

Comets, Meteors and Interplanetary Space

The word *comet* no longer brings with it a flavour of fear or evil. We have been able to find out several things about comets, such as their paths, their composition, and their effect or otherwise on the earth's atmosphere.

We have already shown in Figure 1.1 comet Halley when it passed near the earth in 1985-86. Figure 17.1 is a photograph of another comet, comet Arend-Roland. Colour Plate 20 is a photograph of comet Ikeya-Seki.

WHAT ARE COMETS?

Comets are small bodies in the solar system which orbit the sun, much like the asteroids. There are two major differences between objects of the two categories: (a) The orbits of comets are highly elliptical, and (b) their orbital planes are inclined to the plane of the ecliptic at all possible angles.

STRUCTURE AND CONTENTS

Comets consist mainly of three parts, a *nucleus* or *head,* a *coma* and a *tail.* The size of the nucleus ranges from as small as 1 km to about 100 km. The coma has a size of about 10^4 to 10^5 km. When a comet is near the sun, its tail may extend to a distance of 10^8 km, which is about two-thirds of the distance of the earth from the sun.

Almost all comets have been found to contain various ices (materials in the condensed form), water, ammonia, methane, carbon dioxide, and other carbon and nitrogen compounds. As the comet comes near the sun in its highly elliptical path, the ices get heated and are vaporized.

There is a constant radiation coming from the sun, with its effective *radiation pressure.* Although it does not have much effect on material in gaseous and liquid form, it pushes solid particles away from the sun. Quite a good fraction of the material of a comet is in the solid as well as gaseous form when it is near the sun. This gives rise to the spectacular features of coma and tail.

It is then clear that a comet will have almost no tail when it is far away from the sun. The tail starts forming when the comet is near the orbit of Mars. It is, of course, the longest when the comet is at its perihelion.

COMETARY PERIODS

Encke's comet has the shortest period of revolution around the sun among all the known comets. Its period is 3.3 years. On the higher side, comets have been found with periods of a few hundred years.

Figure 17.1 Comet Arend-Roland. (*Courtesy:* LO)

Comets are divided into two categories on the basis of their periods. Comets with a period of revolution around the sun of less than 200 years are called *short-period comets* while those with a period of over 200 years are called *long-period comets*.

The orbits of short-period comets generally have a small inclination to the plane of the ecliptic. Their motion is also found to be direct. The orbits of the long-period comets have on an average a large inclination to the plane of the ecliptic. This means they approach the sun from almost any direction (see the cometary orbits in Figure 1.2). Also they show an equal proportion of direct and retrograde motion.

DISCOVERY AND NOMENCLATURE

Suppose a comet is spotted in the sky. Since comets may have small as well as long periods, the first question is whether it is a new comet or the reappearance of one which has been observed earlier.

Until the eighteenth century, in fact, each appearance of a comet was thought to be an individual and isolated phenomenon.

Edmund Halley, for the first time, employed Newton's laws and studied the records of 24 comets of the past. He observed that the comet which appeared in 1682 (during his life-time) had an orbit similar to two comets which appeared in 1607 and 1531. He correctly thought that it was the reappearance of the same comet. He predicted that this comet would return in 1758.

Although this prediction came true, Edmund Halley was not alive to watch the event. The comet was named after him.

Nowadays, new comets are discovered every now and then, more or less like asteroids. For example, 73 new comets were discovered during the period 1956-73; in the same period, 93 reappearances of comets were confirmed.

When a new comet is spotted, a temporary name is given to it. Again the style is similar to the case of asteroids. This name consists of the year of discovery followed by a letter of the alphabet. For example, 1968d refers to the fourth comet discovered in 1968. A permanent name is given to it only after its perihelion passage, that is, after it has passed the point on its path nearest the sun.

COMETARY ORBITS

The smallest perihelion distance of a comet has been found to be 0.005 AU. There is no upper limit to the aphelion distance of comets as some of them go far beyond the last planet, Pluto.

Some comets may have orbits other than elliptical orbits, such as *parabolic* or *hyperbolic* orbits. Not all the comets make reappearances. Only comets having elliptical orbits return again and again. Those with parabolic and hyperbolic orbits make a pass at the sun and then fly away to infinity. They leave the solar system altogether.

Comets are small bodies and their paths get highly disturbed when a comet passes a massive planet. Every time a comet passes near the sun, it loses some of its mass in the form of dust and vapour. Even those comets that come back *do not return along the same path*. Their orbital parameters show a noticeable change during successive cycles.

It may happen that a comet is seen a few times, and is then perturbed into a hyperbolic orbit and never returns.

METEOROIDS AND METEORITES

Meteoroids are bodies generally smaller than comets and travelling through the solar system along almost random paths in all possible directions. When the path of a meteoroid intersects that of the earth in transit, it enters the earth's atmosphere. Here its temperature rises due to its high velocity and friction with the atmospheric gases. The meteoroid melts and even starts burning. In this burning phase, a meteoroid is called a *meteor.*

This gives us the appearance of a *shooting star* or, according to the common practice in India, a *broken star.*

A careful observer may easily see five to ten meteors per hour during a night watch. It is estimated that the number of meteors which become visible is about 24 million per day. The total number of meteoroids which enter the earth's atmosphere but are not visible runs into several thousand million every day.

The meteors that we see as shooting stars are meteoroids of size about a millimetre to a centimetre. When its size is larger than a few centimetres, the meteor appears as a huge fire ball, which might hit the earth before it is burnt out. So what we receive on the earth is a *meteorite* which, by definition, is the residue of a meteor.

Several observers have watched and recorded meteors throughout the night and over the years. The following two conclusions can easily be drawn on the basis of these observations:

(a) More meteors are seen *after* midnight than *before* midnight. This means that one sees more meteors during the second half of the night than the first half.

(b) There are certain periods of the year when one sees more meteors than on an average day. Some such periods are April 19-20, August 10-11, November 11-14, etc.

The former observation is explained on the basis of the earth's rotation about itself and the assumption that meteoroids wander in the outer space randomly with all possible speeds in all possible directions. Figure 17.2 shows the earth hurtling in its orbit around the sun at a speed of 30 km/s (to the left in the figure). Sunrise would be taking place at the point R, sunset at S, and there would be midnight at M. The portion marked II in this figure will then encounter all meteoroids moving towards it (to the right) plus those moving to the left with a speed *less than* 30 km/s. On the other hand, only

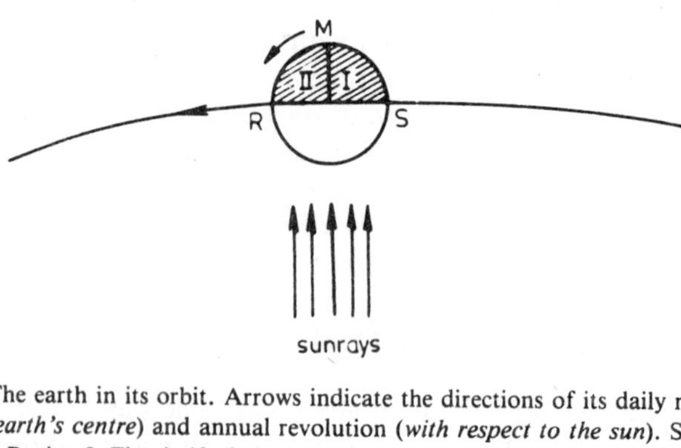

Figure 17.2 The earth in its orbit. Arrows indicate the directions of its daily rotation (*with respect to the earth's centre*) and annual revolution (*with respect to the sun*). Shaded region indicates night. Region I: First half of night; Region II: Second half of night. Region II would encounter more meteors than Region I.

meteoroids moving to the left with a speed *greater than* 30 km/s would enter the earth's atmosphere in region I.

The second observation has led to a suggestion that there are distributed debris of comets all along their orbits and moving along the *same* orbit, one behind the other, like a chain. The orbit intersects the earth's orbit at a particular point and the earth arrives at this point at a fixed time of the year. The number of meteors entering the earth's atmosphere shows a considerable increase during such periods of the year and the phenomenon is known as *a meteor shower.*

LIMITS OF THE SOLAR SYSTEM

It is guessed that there may be several comets in the solar system with extremely long periods. A comet with an extremely long period of revolution around the sun would go to very large distances away from it. In fact, such a comet would spend most of its time in parts of the orbit away from the sun and would not be visible to us during these periods.

Where can we say is the limt of the solar system?

The outermost planet that we have been able to discover is at an average distance of less than 40 AU from the sun. Then there is a postulated cloud of comets, called the *Oort cloud,* at a distance of about 20 000 to 90 000 AU. The distance between the sun and the nearest star is 266 700 AU and the average distance between two stars in our galaxy is even larger than this.

There is no region in the galaxy which is totally free from the gravitational pull of any star. Therefore, we can roughly take the midpoint of the line joining the sun to the nearest star Proxima Centauri as the limit of the solar system. This gives us 2×10^{13} km as the radius of the sphere inside which bodies are influenced by the gravitational field of the sun. However, the escape velocity of a body at such a great distance is about 120 m/s, which is too small for the natural motions of interstellar particles.

There are estimates that there could be millions of comets lying at far distances from us, inside this sphere of influence.

INTERPLANETARY SPACE

You must realize that in addition to the observable large bodies, the solar system contains innumerable small bodies which do not come directly under our observation. We have seen in Chapter 15 that we now know 4865 asteroids. One can only imagine how many more asteroids there might be of smaller dimensions roaming around in the solar system!

Similarly, we have said above that several hundred thousand meteors enter the earth's atmosphere every day. The number of such bodies moving throughout the solar system in all possible directions is really staggering!

Finally there are small particles, molecules, atoms, subatomic particles such as protons, electrons and others, and radiation, filling the entire space. *The interplanetary space is not empty.*

The one thing that you can safely conclude is that, in addition to planets, satellites, asteroids and comets, the solar system consists of numerous bodies of all sizes, from the sun down to subatomic particles and radiation. In the interplanetary space, it is a question of the *density* of matter, which is roughly the number of particles per unit volume.

In Box 9.2, we have told you that there are about 10^{13} or 10^{14} molecules of matter per cubic metre on the surface of the moon at 20°C. The density of matter in the interplanetary space is at least a million times smaller than that on the moon's surface. This means that there would be about 10^7 to 10^8 atoms or molecules of matter per cubic metre in interplanetary space.

18

Exploration of the Solar System

How does one go about exploring the solar system?

Obviously the best thing to do would be to go to the place we wish to study, may it be a planet, its satellite, or comet. That gives us one method: sending manned spacecraft to various destinations.

Manned flights have so far been undertaken only to the moon. Such flights are extremely risky, difficult and costly to operate, and involve delicate manoeuvres. During the next decade or two, there will be efforts to send manned spacecraft to Mars and its satellites.

Manned flights to other planets will remain a far cry for at least several decades. In the case of Venus and Jupiter, our spacecraft and spacemen will have to withstand pressures 100 times or more than that on the earth due to their thick atmospheres. And what about the sun? We cannot go even near it for the fear of being burnt.

The next best thing to do would be to send unmanned spacecraft. Such spacecraft could carry scientific instruments and various experiments to be carried out in the outer space. Such spacecraft could again be of two categories: those that can be brought back to the earth and those that will go away forever. Needless to say, spacecraft of the former category would give us much more information than those of the latter category.

The Soviet Union did send an unmanned spacecraft named Venera to Venus. It slowly descended the thick Venusian atmosphere with the help of a parachute and radioed a vast amount of information and photographs of the planet. Several countries sent unmanned spacecraft which flew past the Halley comet in 1985-86 and studied it from close quarters.

The USA sent two unmanned Voyager spacecraft in 1977 to visit the outer planets. While Voyager 1 visited Jupiter and Saturn, Voyager 2 visited Jupiter, Saturn, Uranus and Neptune. The trajectories of both the spacecraft with dates of launch and dates of close passes at the various planets are shown in Figure 18.1. We have already seen in the previous chapters that these spacecraft have discovered many more moons of these planets and radioed a vast amount of information back to the earth. Both of them are now lost in the cosmos forever.

But we have really begun in a lopsided manner. The best or ideal things are not easy to come by. Such efforts would take decades and centuries.

We cannot wait all that long in our quest of the solar system. Moreover, we must have a fairly good knowledge of the planet or the satellite where

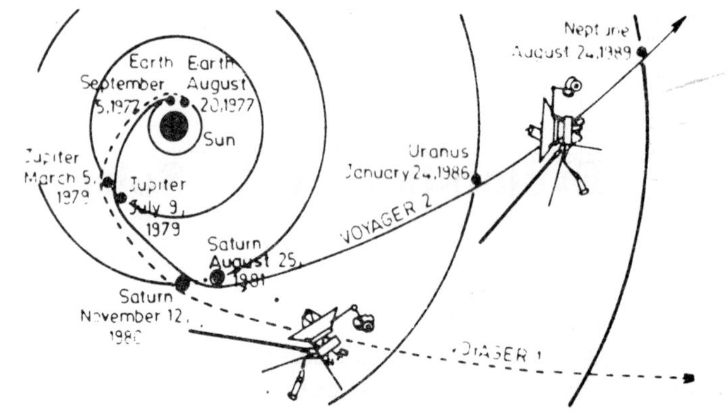

Figure 18.1 The flight trajectories of Voyager 1 and Voyager 2 which visited the outer planets. The launch dates and the fly-by dates at various planets are also shown. (*Courtesy: NASA*).

Box 18.1 *Galileo spacecraft to Jupiter*

The National Aeronautics and Space Administration of the USA has launched a spacecraft, named Galileo, on October 18, 1989, towards Jupiter. Galileo will travel 4000 million km on a convoluted path that will first take it near Venus, then twice past the earth before swerving towards Jupiter. It will take photos and gather data on Venus and the earth-moon system during these passes.

En route to Jupiter, it will make the first ever close-up studies of an asteroid. Figure 18.2 shows the flight path of Galileo.

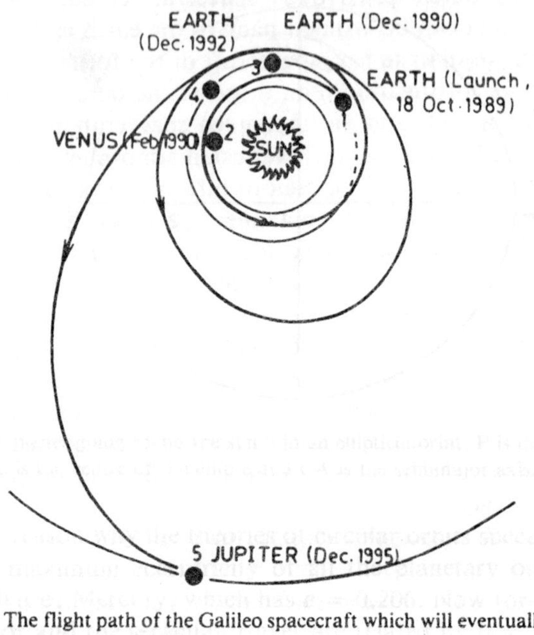

Figure 18.2 The flight path of the Galileo spacecraft which will eventually go into an orbit around Jupiter. (*Courtesy*: SPAN)

When it nears Jupiter, it will release a probe which will enter the planet's thick atmosphere while the main Galileo craft enters into an orbit around Jupiter. The probe's velocity will be slowed down by a parachute. Before the probe perishes under the planet's massive pressures, it will transmit a good deal of data to the Galileo orbiter which will in turn relay them to the earth.

The main Galileo will itself spend 22 months around Jupiter studying the planet and its 16 satellites. Scientists expect to get information which will have direct bearing not only on the origin of the solar system but also on the origin of the universe.

we wish to send manned or unmanned spacecraft, through other methods of observation, before planning a flight to that destination.

Historically, until a few centuries ago, we had only the eye for observing the sun and the planets. Then came the optical telescope, followed by the camera. Now we have a whole battery of technological gadgets and scientific methods of observation and interpretation for this task. We shall consider some of these here.

THE OPTICAL TELESCOPE

This instrument still remains one of the most popular gadgets of observation. We have earlier told you (see Chapter 8) that a telescope is known by the size of its objective lens. Telescopes of a 10-cm or 15-cm objective are quite common in science departments of schools, colleges and universities, and even in science museums and science clubs.

In India, we have major telescopes at several places including Naini Tal, Udaipur, Hyderabad, Kavalur and Bangalore. These telescopes have objectives of sizes ranging from 40 cm to 234 cm. Scientists of the Physical Research Laboratory at Ahmedabad are building a new telescope at Mount Abu, a hill station in Rajasthan.

Among the large optical telescopes in the world, the 508-cm telescope at Mount Palomer is famous. It has been the largest telescope in the world, next only to the Soviet telescope of 610 cm. There is news that the USSR and some European countries are planning to build bigger telescopes.

The European Southern Observatories have recently commissioned the *New Technology Telescope* in October 1989 (aperture size 3.6 metre). This has been put up in Chile in South America. With the use of the new techniques of adaptive optics, it gives a resolution of 0.18 arcsec, which is the best among all the ground-based optical telescopes today.

Various filters can be used in conjunction with telescopes to decipher the complicated radiation coming from the sun and other stellar objects. This amounts to focussing on a particular 'window' for screening the information coming from the object and suppressing the rest of it. Various such windows can give us complimentary information. We then have to build the total picture of the stellar object that is millions and billions of kilometres away from us.

Figure 18.3 is a photograph of the sun obtained by using a calcium K-line filter. This filter allows a band of light near a wavelength of 3933 Å which lies just beyond the violet end of the spectrum and at the beginning of the ultraviolet region. You can compare it with Figure 8.3 which shows the sun photographed in white light.

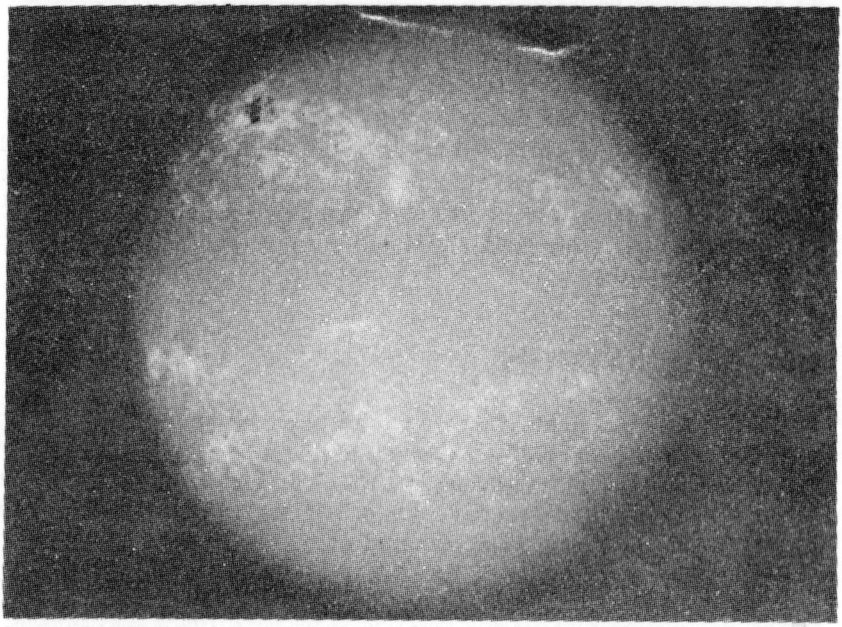

Figure 18.3 Sunspots and calcium plages photographed through calcium K-line filter using 15 cm Solar Coude refractor; March 9, 1989. (*Courtesy*: UPSO)

SPECTROGRAPHS

You are surely aware of the fact that when white light passes through a prism, it splits up into a *spectrum* of colours. Like a rainbow, the visible spectrum stretches from violet at one end to red at the other.

Each atom in the universe, of any element, such as hydrogen, calcium, iron or any other, gives out radiation of certain specific wavelengths, when excited. These wavelengths act like signatures or fingerprints of that particular element.

The spectrum of radiation coming from the sun, any planet, comet or such object gives us information about the elements that are contained in it. This is also true of a star.

Tremendous scientific and technological advances have been made in the recent past to analyse the radiation coming from a stellar object. One such

Box 18.2 *A telescope in the earth's orbit*

A telescope has been launched in the earth's orbit in April 1990. The telescope, which has been built by the US National Aeronautics and Space Administration, has been named the *Hubble Space Telescope* (HST) after the astronomer Edwin P Hubble.

Telescopes on the earth have a common, serious drawback—radiation coming from space has to pass through the entire atmosphere of the earth. This has two effects. It decreases the intensity of the radiation and distorts the information coming in from space. A telescope orbiting the earth at a height of a few hundred kilometres will be free from these drawbacks and will therefore be better than any telescope on the earth.

The HST will be able to 'see' objects seven times farther into space and 50 times fainter than anything observed by ground-based telescopes. Its resolution will be equivalent to reading a car plate number at a distance of 160 km (which is about 0.07 arcsec). It will greatly help observation of the solar system as well as stars, galaxies and the universe at large.

Needless to say, it will discover many more satellites and asteroids in the solar system.

With this telescope and similar incredible new tools of observation and discovery, scientists believe that the 1990s will be the golden era of astronomy.

instrument is called a *spectrograph*. A spectrograph specially built for the study of the sun is called a *spectroheliograph*.

Figure 18.4 shows the low-dispersion spectra of two comets, comet Arend-Roland and comet Mrkos. The existence of the carbon molecule and other carbon compounds has been confirmed in the material of these comets from such spectra.

Box 18.3 *The electromagnetic spectrum*

We now know that the visible region is only a small window in the complete *electromagnetic spectrum*. At the one end of this spectrum, there are very small wavelengths, of the order of an angström or less, which fall in the X-ray or gamma ray region. At the other end, there are long wavelengths, which themselves stretch from a centimetre to a kilometre, and fall in the *radio wave* region. The entire spectrum thus provides us several windows to look at the universe.

Figure 18.5 shows a spectrum of Venus obtained in the infrared region. It betrays the existence of carbon dioxide in the atmosphere of Venus.

RADIO TELESCOPES

A new armour in the battery of an astrophysicist is the *radio telescope*. The radio window is extremely useful in spotting the presence of various atoms and molecules in various stellar bodies.

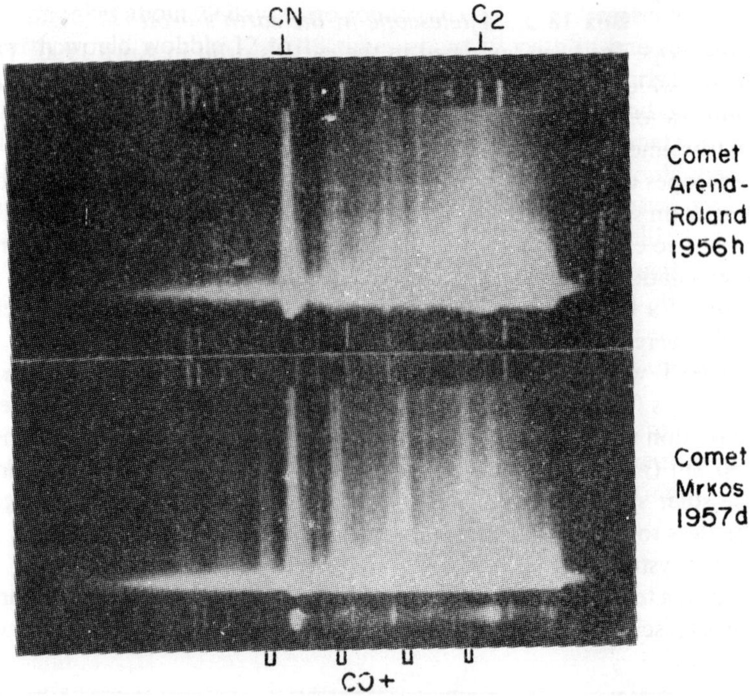

Figure 18.4 Low-dispersion spectra of two comets showing the lines of various carbon compounds. Top: Comet Arend-Roland, 1956h; Bottom: Comet Mrkos, 1957d. (*Courtesy*: LO)

The first radio telescope was built in 1932 by Karl Jansky.

The first radio telescope in India was built in 1971 by the Tata Institute of Fundamental Research (TIFR) at Udhagamandalam in Tamil Nadu. The centre established by the TIFR here is known as the Radio Astronomy Centre. This 530-m long and 30-m wide steerable radio telescope has been successfully used to study bodies in the solar system as well as stars and galaxies in the universe. Figure 18.6 shows this radio telescope.

The National Radio Astronomy Observatory in Charlottesville, USA, has a Very Large Array (VLA) radio telescope. It consists of an array of 27 antennas, each of which is a 25-m reflector antenna. They are arranged along three equally-spaced radial arms, forming a Y-shaped array. Two of the arms are 21 km long while the third, the northern branch, is 19 km long. Colour Plate 21 is a photograph showing the positions of these antennas.

Colour Plates 22, 23 and 24 show how we can get complementary information from pictures taken with radio waves. Colour Plate 22 is a photograph of the sun in radio waves showing its brightness distribution. Each bright coloured spot in this picture is an active region in the sun's corona, above a group of sunspots. High-density hot material is confined in these regions by strong magnetic fields. It also shows that the spots are symmetrically placed about the sun's equator.

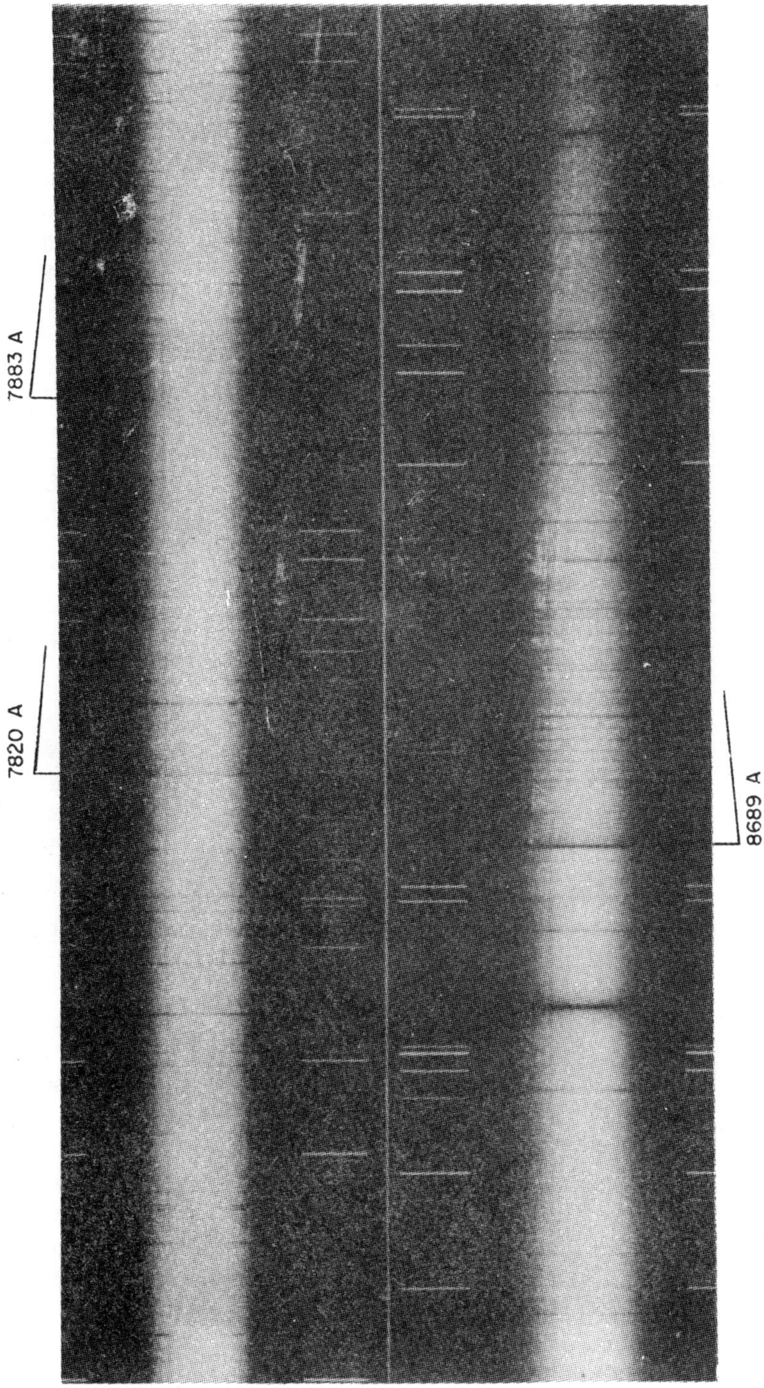

Figure 18.5 Infrared bands of Venus, showing the carbon dioxide lines. (*Courtesy*: LO)

Figure 18.6 The antenna array of the radio telescope at the Radio Astronomy Centre at Udhagamandalam. (*Courtesy:* TIFR)

Colour Plate 23 is a picture of Jupiter obtained in radio wavelength of 21 cm. The extended features beyond the planet's disc (which is the central region in the picture) betray the existence of a magnetic field with charged particles trapped in it. These features could not have been seen with an optical telescope. Jupiter has a strong magnetic field which traps fast moving electrons, which in turn emit radio waves and give rise to these features.

Colour Plate 24 shows how Saturn looks in radio waves, with a wavelength of 2 cm. Saturn's atmosphere is clearly seen in radio waves, surrounding the planet's disc. Its rings are also seen outside the planet's disc, but appear as a dark band in front of the planet.

Box 18.4 *National image processing facility in astronomy*

Nowadays, you don't look at the planets and stars in the sky directly or even through old-fashioned telescopes (that is, if you are a scientist); you look at their images obtained through sophisticated instruments, processed by computers, and displayed on video screens in your laboratory.

A National Image Processing Facility in Astronomy (NIPFA) was set up at the Radio Astronomy Centre, Udhagamandalam. It became fully operational in early 1986 and has since been used with commendable efficiency.

It offers excellent hardware and sophisticated techniques for data processing and interactive image processing. In addition to radio astronomy, it is also being used significantly for optical and X-ray data.

GIANT METREWAVE RADIO TELESCOPE

A Giant Metrewave Radio Telescope (GMRT) is being set up by the National Centre for Radio Astrophysics (NCRA) of the Tata Institute of Fundamental Research as a National Facility at Khodad, near Narayangaon, 80 km north of Pune. The academic headquarters of the NCRA are located on the campus of the University of Poona.

When completed in 1992, the GMRT will become the most powerful facility in the world for astronomical research near the metrewave-band. It is being designed to investigate a wide variety of celestial objects, ranging from our solar system to the very edge of the observable universe.

Other radio telescopes in the world have so far explored the universe through wave-bands around a centimetre. But there are many unexplored, outstanding astrophysical problems which can be studied only near the metrewave bands and the GMRT has been specifically designed to investigate these.

India is the only country among the developing ones which has the requisite technical know-how and capacity to build such a telescope *and* where a vast area of about 30 km × 30 km can be found which is reasonably free from industrial and electronic noise. The site of the GMRT is situated in a valley surrounded by mountains in the Western Ghat range.

This radio telescope will consist of 30 parabolic dish antennas, each of 45-m diameter. Twelve of these will be placed in a central pattern in an area slightly larger than 1×1 km^2, and six each will lie along three 14-km long arms; see Figure 18.7. This arrangement is known as the Y-shaped array. At the position of each dot in this figure, there will be a 45-m diameter parabolic dish antenna, hoisted 20 m above the ground. Figure 18.8 shows a sketch of one such antenna. The four rods at the top constitute what is known as the *quadrupod.*

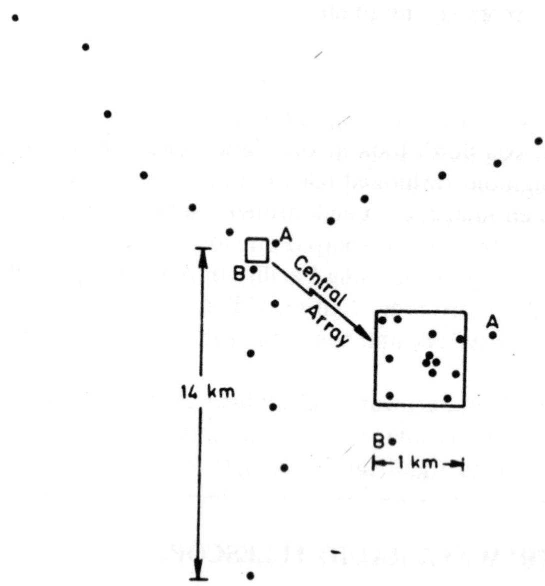

Figure 18.7 Schematic diagram of the locations of the 30 antennas of the GMRT. (*Courtesy*: NCRA, TIFR)

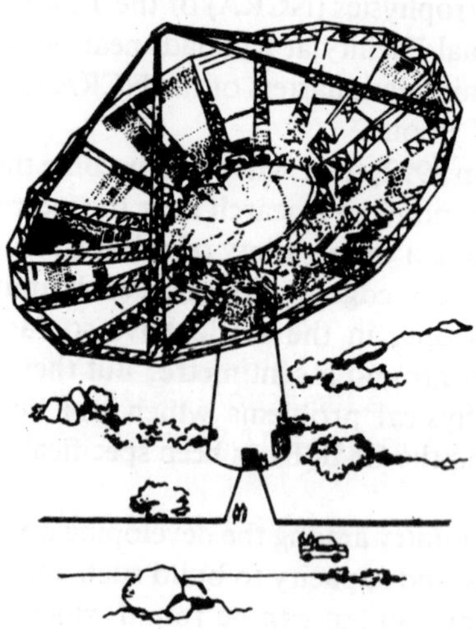

Figure 18.8 A sketch of one of the dish antennas of the GMRT hoisted on its tower. (*Courtesy:* NCRA, TIFR)

Colour Plate 25 shows the dish antenna on the ground and without the quadrupod, during the period of the assembly of the telescope.

The system will operate in such a manner as to give a resolution of 2″ at the frequency of 1420 MHz (which corresponds to a wavelength of 21 cm), coupled with a high sensitivity.

One of the proposed projects GMRT will undertake is high-resolution imaging of the sun, in addition to the imaging of galactic and extragalactic radio sources.

DETECTION OF GRAVITATIONAL WAVES

An Inter-University Centre for Astronomy and Astrophysics (IUCAA) has been set up by the University Grants Commission as a centre of excellence in astronomy and astrophysics on the Pune University Campus in 1988. One of the main observational programmes undertaken by this Centre is the detection of gravitational waves emitted by supernovae and binary pulsars. An Astronomy Data Centre and an Image Processing Facility are also being set up at IUCAA.

19

Place of the Solar System in the Universe

HOW BIG IS THE SOLAR SYSTEM?

When we say that our earth has a radius of about 6370 km and a mass of about 6×10^{24} kg, we hardly appreciate its bigness. Should we state the same facts in the following way, it will certainly surprise many of you. Suppose the entire earth is cut into pieces and loaded in wagons each of capacity 20 tons and of length 5 metres. The distance between the front and the rear ends of the supposed train will be 1.5×10^{18} km. This comes to about 160 000 light-years, which means that the green signal sent by the guard at the rear will reach the engine driver at the front only after 160 000 years. So big is our earth!

Now let us represent the size of our earth by that of a marble of radius 6.4 mm, then the size of the moon on this scale will become a peanut of radius 1.7 mm, and this peanut will have to revolve around the marble (earth) at a distance of 38.4 cm. If we maintain the same scale as above, we shall then have to place our sun at a distance of 150 metres. The sun itself will have a radius of 70 cm. The farthest planet Pluto will be situated at a distance of about 6 km and its size will be no bigger than a mustard seed.

What a big void is the solar system! And how big it is too!

We then plan to leave our solar system and look for the nearest star. The nearest star other than the sun is situated in the constellation Centaurus. When both Virgo and Scorpio are seen high up in the middle of the sky (at about 8 pm local time in June), the constellation Centaurus appears near the southern horizon with two very bright stars in it (just to the left or east of the famous Southern Cross), named beta and alpha Centauri, or Hadar and Rigil Kentaurus, respectively (see Figure 19.1).

If you look through a 10-cm telescope with a magnification of, say, 40, and focus at alpha Centauri, the brighter one of the above two, you would find a bright companion of this star. These two components are named alpha Cen A and alpha Cen B. The latter is slightly fainter than the former but brighter than beta Centauri. Both these stars are at a distance of 4.35 light-years from us.

However, the alpha Centauri system actually consists of three stars, the third one being extremely faint. It is the one *closest* to us. For obvious reasons, it is named Proxima Centauri (Proxima meaning nearest). Its distance from us is about 4.22 light-years. It is an eleventh magnitude star to be seen about

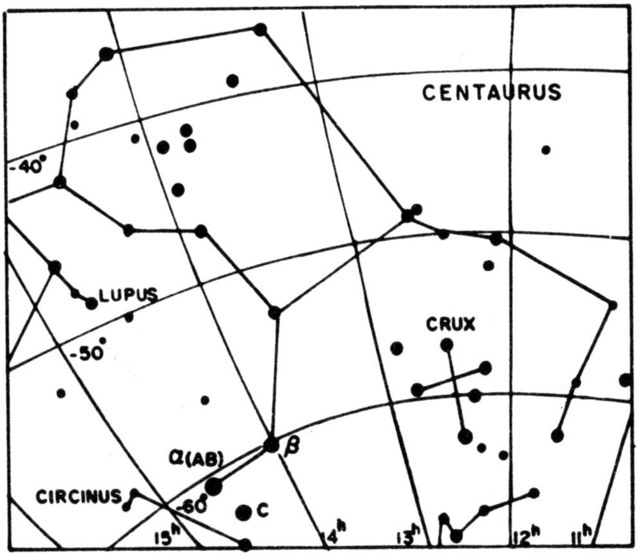

Figure 19.1 The entire constellation Centaurus is shown with the Southern Cross (technically known as *Crux*). To the left (east) of Crux there are two very bright stars, β and α Centauri, α being brighter than β. Through a telescope, α is seen as a double star, the components named as A and B, A being brighter than B. About 2° south of the α (AB) there is a component star C which is extremely faint (11th magnitude), to be seen only through a 20-cm or bigger telescope.

2 degrees of arc away from the Cen AB system, the separation between A and B being only 19″ at present.

Thus the light which you receive from Proxima Centauri must have left the star about 4.22 years ago. Remember that it takes about 5.5 hours for light to come from Pluto, and only 500 seconds from the sun.

However, if we now use our above-mentioned scale of the model solar system with the marble representing the earth, Proxima Centauri has to be placed at a distance of approximately twice the distance between Bombay and California! So stars are like flourescent balls floating in space, say one in Calcutta, another in London or Tokyo or Sydeny, and the planets around each of the stars are like a few flies roaming around these lighted balls within about a kilometre or so from the respective balls.

THE SOLAR NEIGHBOURHOOD

The *solar neighbourhood* is defined to be the region of space in which there are about a million stars centred about the sun. You may think that it would surely occupy a large portion of the universe. Not at all! Its radius is only about 330 light-years.

Well, first to introduce you to some nearby stars, we can certainly choose our *dog star* Sirius, apparently the brightest star in the sky and to be found a few degrees south-east of Orion, the *hunter* constellation. Sirius is only about 8.6 light-years away from us. But consider Canopus, the next brightest star in the sky, to be found in the constellation Carina. If you see Sirius high up in the sky, look further down to the south and there is Canopus awaiting your attention. Its distance from us is about 74 light-years.

Now look at the constellation Orion. Its left shoulder and the right foot consist of two very bright stars named Betelgeuse and Rigel, respectively (see Figure 19.2). Can you guess how far they are from us?

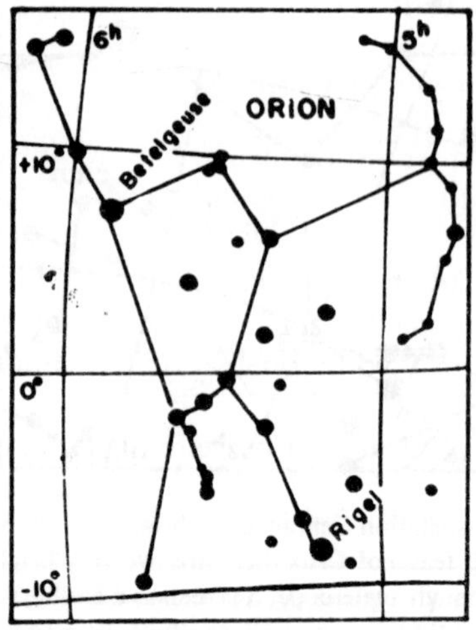

Figure 19.2 The constellation Orion is shown along with the two bright stars Betelgeuse and Rigel.

In fact, most bright stars in the constellation Orion are about 310–1400 light-years away! So the light of Betelgeuse that you see today has left the star when Shivaji was fighting with the Mughals.

Now shift your attention slightly to the east of Orion until Procyon catches your eyes. It appears a little fainter than Betelgeuse but its distance from us is only about 11 light-years. Our pole star is situated about 800 light-years away. It is therefore practically impossible for anyone to guess the distance of a star merely from its appearance. Of all these stars, alpha Centauri is very much like our own sun, in terms of its size, luminosity and surface temperature. So you can easily guess that if we could push our sun to the distance of alpha Centauri, it would look just like alpha Centauri.

But this is not generally true for other stars we have mentioned above. For example, Sirius is intrinsically about 19 times brighter than the sun, Proxima Centauri only about 1/300th of the sun, Canopus about 830 times brighter, Polaris about 9000-times brighter, and Rigel 145 000 times brighter than the sun. So the intrinsic brightness of stars can vary over a very wide range.

From a countryside, under the best conditions of weather, we can see about 8500 stars in all with the naked eye. They come at a variety of distances, ranging from 4.22 light-years to the farthest ones usually at distances of 3000 light-years. However, most of the stars visible to the naked eye are intrinsically very bright, brighter than the sun. We now know that more than 90%

of *all* stas are subluminous compared to the sun (which means that they are less bright than the sun). Therefore, up to the distance of the pole star, the expected total number of stars is in fact about 10 million, of which 99.15% are really so faint that they are not seen with the naked eye.

A bright star like Rigel is very rare, possibly one in ten thousand or less. We in fact see a few of them among the naked eye score of 8500, implying that we are luckily situated near a relatively thick population of bright stars (called the OB associations) and, at the same time, near what is called the Perseus-Orion arm of our Galaxy, the Milky Way. We shall explain in the following section some of the strange objects mentioned here.

INTERSTELLER MEDIUM AND STRUCTURAL PATTERN OF OUR GALAXY

The space between the stars is not totally empty. We know that stars are hot blobs of gas, which means that stars are formed out of gas. Huge clouds of gas condense under their own gravity and disintegrate into several component clouds, which again contract and finally form stars. In the process, not all gas can condense into stars. Some gas is left behind to form what is known as the *interstellar medium* (ISM in short). The following simple calculation can give you an approximate idea of the density of the interstellar medium.

We have seen that the average separation between stars can be 4 to 10 light-years. Let us assume that inside a cubic volume, having each side of 10 light-years, there resides interstellar gas equivalent to one star as a leftover of the star formation. A sun-like typical star has a mass of 2×10^{30} kg. Since the mass of a hydrogen atom is 1.67×10^{-27} kg, the total number of hydrogen-like atoms in a sun-like star comes to about 1.2×10^{57}. Now let us distribute these atoms inside a volume of (10 light-years)^3 = 8×10^{50} m^3, giving the average interstellar density of about 1.5×10^6 hydrogen atoms per cubic metre, which is about one or two hydrogen atoms per cm^3.

In fact, from different observations, it is found that the interstellar medium between stars in the solar neighbourhood roughly corresponds to the value estimated above.

A large fraction of the ISM, about 98%, consists of gas and the remaining 2% is made up of tiny dust grains, the typical size of which is less than 1 μm (10^{-6} m). Of the 98% gas, about 60% is hydrogen and 35% helium, and the rest is in the form of heavier elements like carbon, oxygen, silicon, iron, etc. Of the hydrogen, more than about one-third is made up of molecular hydrogen, the rest being in the form of neutral atomic hydrogen or ionised hydrogen. Usually the ISM around hot stars is totally ionised whereas that around cooler low-mass stars remains almost totally neutral.

Recently a satellite called the *Ultraviolet Explorer* has discovered an astounding fact about the surrounding of the sun up to about 30 light-years from it. The ISM is found to be locally very hot. It emits ultraviolet light and even soft X-rays. The particle density in this region is very low, less than

0.1 hydrogen atom per cm^3. This is called the *local super bubble* of the ISM. The sun is situated almost at one end of this bubble and a few small cloudlets are merely floating inside this bubble. The bubble was caused perhaps by a supernova explosion in the past in a region within 10 to 20 light-years from the sun.

OUR GALAXY: THE MILKY WAY

On a moonless night, one can easily notice a long patch of diffuse light extending from one end of the sky to the other, running through the constellations Scorpius, Sagittarius, Cygnus, Perseus, Taurus, Puppis and so on. It is our *Galaxy,* known as the *Milky Way,* a galaxy of stars in which our solar system is situated. It is made up of an innumerable number of faint stars—faint because of distance. Our Milky Way is just one galaxy among the billions that exist in the universe.

The sun and the other stars are not motionless objects in the universe. In fact, all of them are moving. If we take the entire solar neighbourhood as a collection of a few million stars, we can easily find out how its centroid (or centre of mass) moves. We now know that the centroid of the local system of stars moves around the centre of our Galaxy with a speed of about 220 km/s in the direction of the constellation Cygnus.

Figure 19.3 shows an edge-on view of the Milky Way. A new unit of distance, a *kiloparsec* (symbol kpc), has been introduced in this figure. One parsec (pc) equals 3.26 light-years. The sun is at a distance of 8.5 kpc or 27 700 light-years from the centre of our Galaxy.

Again, within the local system of stars, each star moves with a random velocity about the centroid of the local system. Such a motion is called the *peculiar motion* of an individual star. Our sun moves with a peculiar velocity of about 20 km/s with respect to the centroid in the direction of the constellation Hercules.

Our Galaxy consists of several arms of dense gas and stars, the arms being arranged in a spiral fashion. Each bit of an arm can be considered as a local system. Our sun belongs to the so-called Persius-Orion arm of the Galaxy. As we said above, this arm moves with a speed of 220 km/s about the centre of our Galaxy, which is situated 28 000 light-years in the direction of the constellation Sagittarius. Figure 19.4 shows another view of our Galaxy, with stars scattered in spiral arms. It takes about 240 million years for one revolution of the local system around the centre of the Galaxy.

Our Galaxy contains no less than 600 billion stars (one billion is one thousand million). This number is so large that if all the stars of our Galaxy were equally distributed among the 5.3 billion humans of the earth, each one of us would have more than 100 stars. An average star in the Galaxy is slightly dimmer and less massive compared to the sun. Since stars are continually forming inside big clouds, gas is continually expended to form stars. Only a few of them die, while the rest live happily. So, by now, hardly 15% of the total mass of the galaxy is left out as gas, while the rest has been converted into stars.

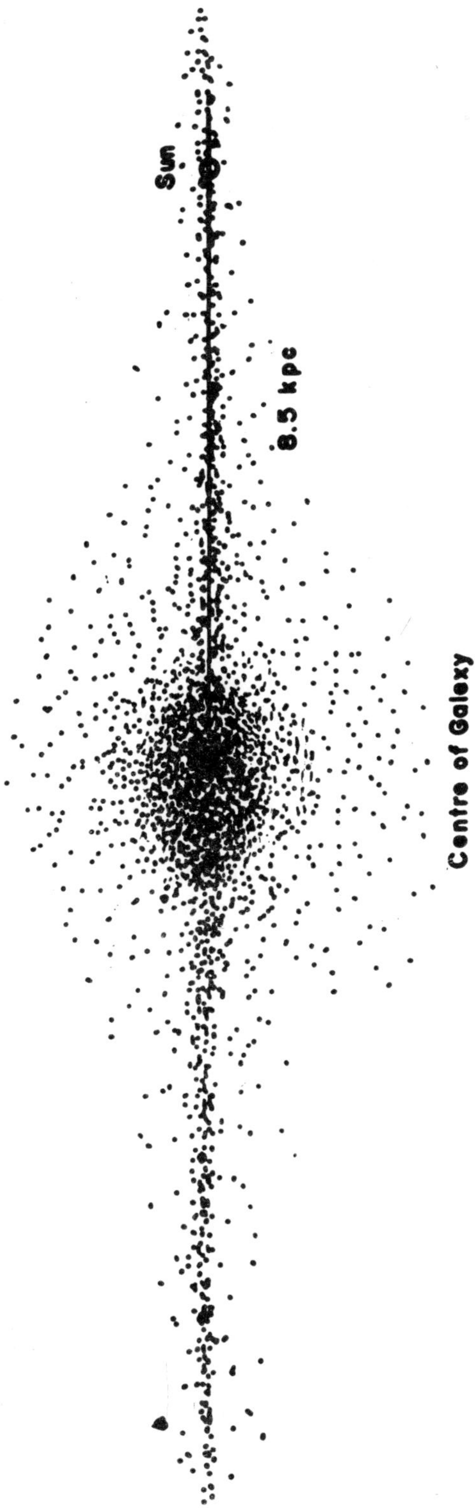

Figure 19.3 A side view of the Milky Way Galaxy. The small circle shown below the word 'Sun' represents the entire solar neighbourhood containing about a million stars.

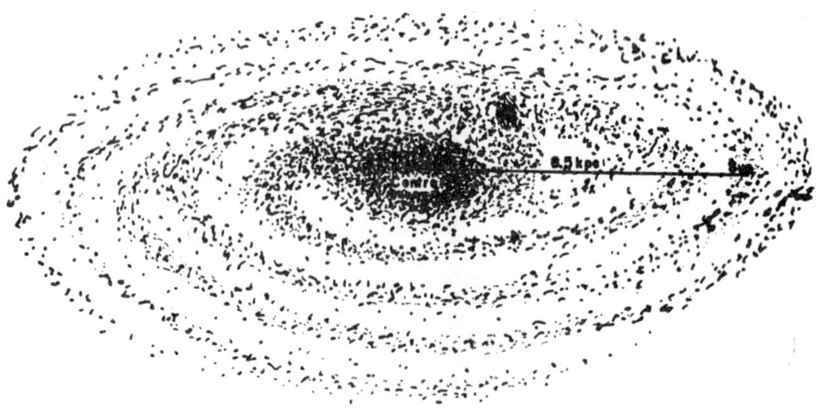

Milky Way Galaxy

Figure 19.4 Top view of the Milky Way, showing the spiral structure. The location of the sun is shown inside one of the arms.

THE UNIVERSE

The universe is very big. There are several billion such galaxies in the whole universe, each containing on an average 200 billion stars. The galaxies come in different shapes and sizes. For example, some are elliptical or lenticular, some are spirals or barred-spirals, and the rest are without any particular shape, called *irregulars*. Many of these galaxies are as quiet as our own Milky Way, though some are extremely violent. Among them are objects called *quasars, blazars, Seyfert galaxies,* etc.

The most distant object that we have seen so far through any telescope lies at a distance of about 15 billion light-years, which means that light takes about 15 billion years to cover this vast distance. So we can directly see objects which are 15 billion years old. Since this is comparable to the age of the universe, we see quite younger parts of the universe in more and more distant shells around us.

One can easily calculate the typical diameter and density of the universe. If we take the radius of the universe equal to 15 billion light-years, the volume of the whole universe would be about 10^{79} m^3. The mass contained in all the galaxies is of the order of 10^{51} kg. This gives an average mass density of the universe to be about 10^{-27} kg/m^3.

This is an exceedingly low density, implying that there is about one hydrogen atom in every cubic metre of space. It is about one-millionth of the local interstellar density of matter (which we saw to be about one hydrogen atom per cm^3). The universe is not only very big in size but also is truly a huge void containing very little matter in it. If all the matter in the universe were pulled together to form a water-like dense substratum, its expanse would not have gone beyond one-third the distance to the nearest star, or about 1.5 light-years!

IS THERE LIFE ELSEWHERE IN THE UNIVERSE?

This is a perennial question that man has asked ever since he found himself confined to the earth. Today, after sending spacecraft to distant parts of the solar system, we know that humans on the earth are the only intelligent creatures habited anywhere in the solar system. Other planets of the solar system are either too hot or too cold for sustaining life, and too arid or too lethal in atmospheric composition to harbour life over billions of years. Even if we implant life in these alien conditions, it will not survive for more than a few gestation periods, let alone periods extending over several billion years.

Since our Galaxy contains about 10^{11} sun-like stars and the whole universe is made up of several billions of such galaxies, it would be quite natural to think that many stars can have planetary systems around them. In that case, at least some of them would have an earth-like planet with the right conditions for harbouring life. A famous American scientist named Frank Drake, in 1961, devised a formula by which he could estimate the statistical probability of life around any star in general. Most people, along with Drake, hope that we the humans on this earth are not terribly alone.

However, we must remember that more than 80% of all stars are either binaries or have multiple components. Such systems cannot have an earth-like planet revolving for billions of years in a stable orbit which must have to be in the temperate zone where water can remain in the liquid state. In any case, the number of single stars is possibly no more than 20% of all the stars. Drake's equation still provides us a good hope of discovering life elsewhere in the Galaxy.

In the late seventies, the Americans had proposed a very expensive project called CYCLOPS, which then projected a cost of about 5 billion US dollars. It consisted of an array of about 1000 dishes or radio receivers, each of 100 metre radius, covering a total area of about 800 km^2, aimed to receive any intelligent radio message sent by any intelligent life in nearby stars. But the US government could not afford its tremendous cost and did not sanction the project. Other serious attempts have also been made to detect signatures of life in nearby stars using optical and infrared observations. The most notable one is a programme called the *Search for Extraterrestrial Intelligence* (SETI) which plans to scan all directions of the sky in a wide range of frequencies for receiving suitable *intelligent* signals. There are also additional programmes for the record for extraterrestrial life, run by several groups of scientists in the United States, Australia, and the Soviet Union. Particularly notable ones are the Microwave Observing Project of NASA, and a project named SERENDIP of the University of California, Berkeley.

However, nothing conclusive has yet emerged from all these searches for extraterrestrial life.

20

Open Questions

This book would be incomplete without mentioning some important questions and puzzles haunting the physicists and the astrophysicists. Here we discuss half a dozen problems on which current researches are focussing (Figure 20.1).

Figure 20.1 An astrophysicist in quagmire. (Drawing by Sham Deshpande)

THE SOLAR NEUTRINO PUZZLE

We know that the sun produces energy by fusing hydrogen into helium in its core region. Since the nucleus of a helium atom contains two neutrons and two protons, essentially four hydrogen atoms or protons are to be combined to form the nucleus of a helium atom. This means that in the process, two protons have to be effectively converted into neutrons.

Neutrons are almost identical to protons in mass but, unlike protons, are neutral in electrical charge. Hence the nuclear reaction that transmutes a proton into a neutron must accompany the formation of a *positron* in order to balance the electric charge.

However, it was experimentally found that in such reactions, if only the above particles were taken into account, the total energy and the total momentum were not conserved. An elusive particle, proposed by Wolfgang Pauli, ought to carry away the balance of energy and momentum. Pauli named this elusive particle *neutrino*. These particles were later discovered in the laboratory, proving Pauli's conjecture.

Now if our sun has to produce energy by the above process of nuclear fusion, it ought to emit a large number of neutrinos. Every second about 10^{37} hydrogen atoms in the interior of the sun are converted into helium. So, an equivalent number (actually one-half the number) of neutrinos must be created in the central region of the sun every second. Since the neutrinos hardly interact with ordinary matter, they essentially leak out from the central region to the surface of the sun. Some of them eventually reach the earth, with an intensity of about 100 million neutrinos per cm^2 per second. So the human body is literally being bombarded constantly with about 100 billion neutrinos every second.

Some clever scientists, including Raymond Davis, Junior, have devised a terrestrial trap for these neutrinos. Davis has in fact been able to detect neutrinos coming from the sun. But it turned out that their number was about one-third of the number expected from the detailed theory of fusion in the sun. This discrepancy is known as the *solar neutrino problem,* haunting the solar physicists for well over the past two decades. Innumerable contrived solutions to the problem have been suggested, but a reasonable one is yet to come.

THE CYCLE OF SUNSPOTS

We have discussed sunspots in Chapter 8. These are dark spots on the bright disc of the sun which can be seen through a telescope and sometimes even with the naked eye. At any time of the year, the sun is never absolutely free from these spots although their number does fluctuate.

It has long been noticed that the number of reasonably big sunspots varies quite systematically with time. In fact, the variation in the number of sunspots follows a periodicity of eleven years. Solar physicists have not yet been able to understand why the sunspots should have such an odd and peculiar periodicity. The most natural periodicity corresponds to the natural solar oscillations, having periods of several minutes only. Eleven years is supposed to be too long a period.

Sunspots are also found to be concentrated near the equatorial belt of the sun (see Colour Plate 22) and they physically drift across the body of the sun. There are many sophisticated theories for the origin of sunspots, but none of them is considered to be very satisfactory.

JUPITER'S RED SPOT

Since the time of Christiaan Huygens, around 1656, two big reddish spots have been seen on the disc of the planet Jupiter. On closer examination, they appear to be atomspheric whirlpools. But there are some puzzling questions about the spots. How could they remain there for hundreds of years?

We know that all the terrestrial cyclones and their analogues are very short-lived phenomena, lasting hardly for a couple of days. If Coriolis force due to planetary rotation is the driving cause of such cyclones, it is difficult to understand why on Jupiter a big cyclone should be maintained almost motionless for over 300 years.

SUPER-RADIATION OF SOME PLANETS

By the term *super-radiation,* we mean radiation output from an object exceeding its total energy input from its surroundings (here the sun). Planets are illuminated by the light of the sun and their warmth is also due to the heat received from the sun. The spacecraft in the Pioneer series had observed that Jupiter and Saturn emit twice the energy they receive from the sun.

Our earth also behaves similarly to some extent, for which we have a ready-made explanation—the radioactive elements that reside inside the earth provide this extra heat. But in the case of Jupiter, we cannot think of such a large amount of radioactive material. This is because the major constituents of Jupiter and Saturn are found to be hydrogen and helium. So what could be the source of this extra radiation in outer planets?

One of the tentative explanations has been the gradual gravitational settling of helium towards the centre of these planets. Helium, being heavier and more inert than hydrogen, can only gravitationally segregate towards the inner core. But this explanation is not entirely satisfactory since a significant part of the energy is being radiated out in the form of Jupiter's own cosmic rays and radio waves.

PLANETARY MAGNETIC FIELDS

It is now known that most planets, except Mercury, Venus and Pluto, have their own magnetic fields which have significant strengths. The earth also has a magnetic field whose strength is about 0.36 gauss.

It has been established that the orientation of the magnetic poles on the earth has flipped innumerable times in the past. The magnetic north pole of the earth changes its polarity within a few thousand years and then continues with the same polarity for about two million years before the next flip occurs. One can guess that the cause of such flips must be intimately connected to the origin of the magnetic field itself.

Those planets which bear strong magnetic fields are also found to rotate quite fast about their own axes. Their core regions are perhaps in fluid state under conditions of high temperature and pressure. If a part of the fluid gets ionised and rotates around some axis, it can produce magnetic fields with a pattern much like that of a bar magnet.

This kind of tentative explanation works fine, as Venus and Mercury are geologically dead objects and rotate extremely slowly around their axes, resulting in very little or no magnetic field. However, the earth experiences frequent reversals of magnetic field. The same is possibly true of Jupiter, Saturn and Uranus. But the cause of the reversal of the magnetic field is not so far understood very well.

ORIGIN OF THE SOLAR SYSTEM

Till now, there are about five dozens of theories of the origin of the solar system. It must be remembered that we are the most privileged ones to have the most direct access to examine the origin of the earth, and yet, very little is understood about it.

The most puzzling aspect of the origin of the solar system is the formation of the so-called inner and outer planets having widely different chemical compositions. The inner planets are mostly metallic and refractory, whereas the outer planets are primarily gaseous, having a composition quite similar to that of the sun. Did the earth and the other inner planets lose all the volatiles? Or did they form by agglomeration of smaller objects called planetesimals?

The vast number of asteroids, comets and meteoroids that are floating around in the solar system may be the remains of the original condensation of the solar nebula. It is not known of what size the original planetoids were formed, how they coalesced into bigger fragments, how the widely different compositional characterstics of the inner and the outer planets were developed in due course of time, and finally how the satellites were formed.

No definite answers to these questions are known as yet.

Index

4

The Motion of the Visible Planets

At this point, we must say that it is interesting to observe the planets in the sky. Watch their motion with time—over a few hours, a few days, and then over months and years.

Observe the motion of the sun or any planet in the sky. Observe the position of these objects against the 'fixed' stars in the sky. Make a note, preferably a written one, of the position of a planet relative to the sun and the stars. Notice the change in the position of the planet: (a) in an hour or two (on the same night), (b) from day to day (at the same time every day), (c) over a period of several days or months, and (d) over a period of years. These observations will tell you a whole lot about the solar system.

THE DAILY MOTION

Let us first consider the gross motion of objects in the sky which hardly takes a few hours. This is part (a) of the observations mentioned above.

Consider, for example, a planet in the sky. When you observe it at a particular moment, suppose it appears to you to be situated in a certain group of stars. If you observe it for an hour or two, you will find that the planet has a motion towards the west *along with all the background stars*. What is really happening is that you, on the earth's surface, are moving towards the east because of the earth's daily rotation about itself. Thus the *diurnal motion* of a planet is a uniform westward motion in which its position with respect to the stars remains practically unchanged.

This is not surprising at all. Don't we observe that every planet or star, including the sun and the moon, rises in the east, moves from the east to the west in about 12 hours, and finally sets in the west? This is the apparent motion of all objects, which is seen because we on the earth are ourselves rotating around the earth's axis once a day from west to east. We cover an angle of 360° (that is, one full circle) in 24 hours, or 15° per hour. In fact, every object in the sky would possess this uniform westward motion.

Once again, we wish to make it clear that the diurnal motion of a planet is not its real motion with respect to the sun or the stars. It is only the apparent motion of a planet which occurs merely because the earth rotates about its own axis once a day.

To study the real motion of a planet, thus, you must forget its diurnal motion altogether. Concentrate only on its motion relative to the sun and the stars. Of course, its shift in the sky with respect to the stars over a day

Plate 1 View of the earth from lunar orbit. The earth is hovering over the bleak and airless lunar horizon. *(Courtesy*: NASA/JPL)

Plate 2 Photograph of the earth taken by the crew of Apollo 17. The photograph shows the complex weather systems in the southern hemisphere and the continent of Africa. (*Courtesy*: NASA/JPL)

Plate 3 A planetary conjunction. Several planets are seen in a small part of the sky. (*Courtesy*: CFC)

Plate 4 The full moon. (*Courtesy*: LO)

Plate 5 Mercury, photographed by Mariner 10 from a distance of 200 000 km. (*Courtesy*: NASA/JPL)

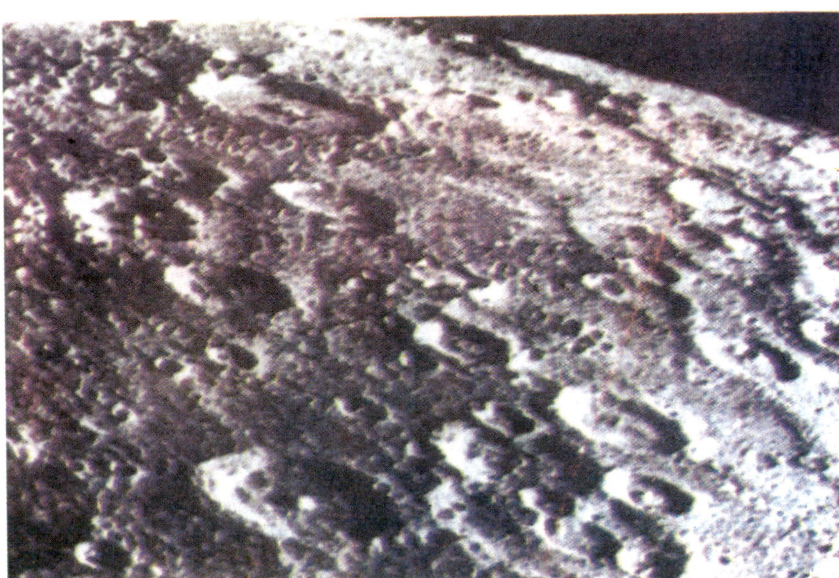

Plate 6 A close-up of part of the north-east quadrant of mercury near the Endeavour Rupes Scarp. The prominent crater with the central peak in the mid-left side of the picture is about 100 km in diameter. (*Courtesy*: NASA/JPL)

Plate 7 Mariner 10 photograph of Venus taken with an ultraviolet filter. The blue colouration is not real but is added to show the circulation pattern of Venusian clouds, which rotate very rapidly. (*Courtesy*: NASA/JPL)

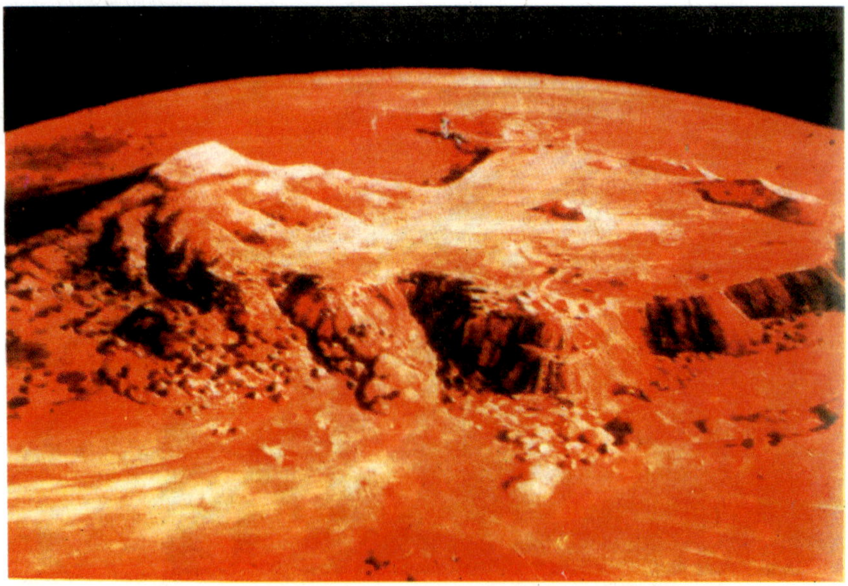

Plate 8 An artist's conception of Ishtar terra, the highest and most dramatic continent-sized highland of Venus, based on measurements by Pioneer Venus Orbiter spacecraft. (*Courtesy*: NASA/JPL)

Plate 9 Photograph of Mars taken by Viking 2 in August 1976, showing dawn along the Tharsis Ridge. (*Courtesy*: NASA/JPL)

Plate 10 Mars photographed by Viking 1 on June 17, 1976, from a distance of 56000 km. Four volcanoes are visible on the Martian surface. (*Courtesy*: NASA/JPL)

Plate 11 Jupiter, photographed by Voyager 1 on January 24, 1979, while still more than 40 million km away from it. The Great Red Spot and the complex pattern of the Jovian atmosphere are clearly visible. (*Courtesy*: NASA/JPL)

Plate 12 Another view of Jupiter with its Great Red Spot. (*Courtesy*: HP)

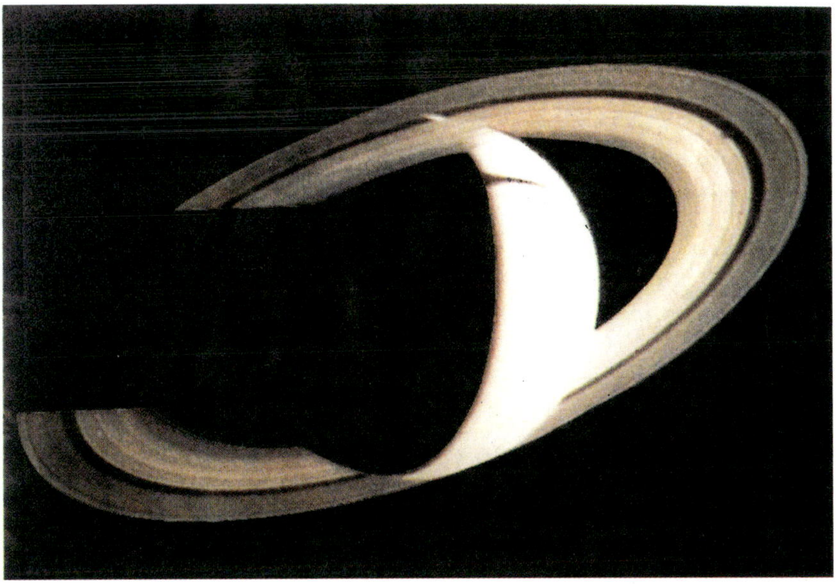

Plate 13 Voyager 1 looks back at Saturn, four days after its fly-by of the planet, from a distance of 5 million km. A crescent view of Saturn, never seen from the earth. The sun is to the right. (*Courtesy*: NASA/JPL)

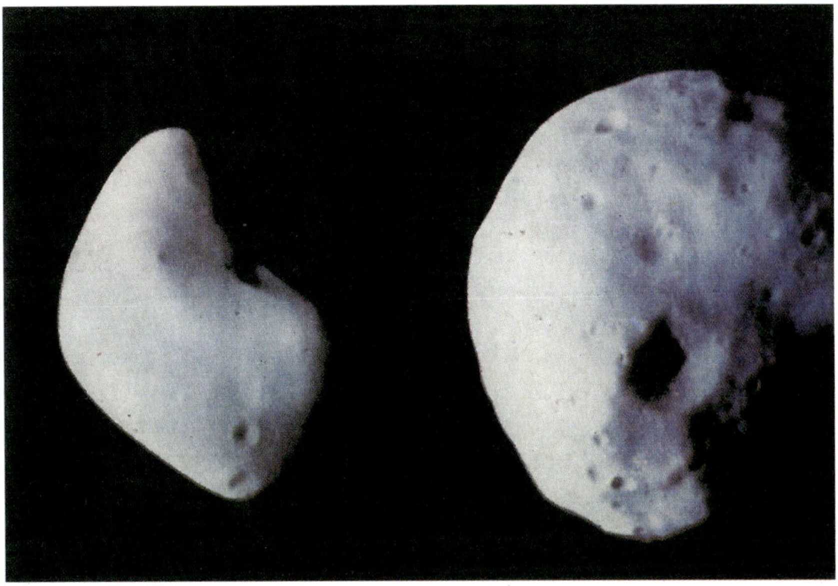

Plate 14 A Viking photograph of the Martian moons Phobos (larger) and Deimos (smaller). (*Courtesy*: NASA/JPL)

Plate 15 Jupiter and its four Galilean satellites; a collage of photographs taken in March 1979 by Voyager 1 (not to scale). Io is at upper left, Europa at centre, Ganymede, and Callisto at lower right. (*Courtesy*: NASA/JPL)

Plate 16 A montage of Saturn and some of its satellites, prepared from photographs taken by Voyager 1 in November 1980. Saturn is at top-left, Dione at bottom-right, Rhea to the left of Dione, Titan at the top, Enceladus to the right of the rings, Mimas closest to the rings, and Tethys in front of Saturn. (*Courtesy*: NASA/JPL)

Plate 17 Pluto and its satellite Charon. A David Hardy painting. (*Courtesy*: NASA/JPL)

Plate 18 An artist's impression of Uranus and its rings. (*Courtesy*: NASA/JPL)

Plate 19 Neptune seen from its satellite Triton. (*Courtesy*: HP)

Plate 20 Comet Ikeya-Seki. (*Courtesy*: CFC)

Plate 21 An aerial view of the Very Large Array (VLA) radio telescope of the National Radio Astronomy Observatory, USA (*Courtesy*: NRAO)

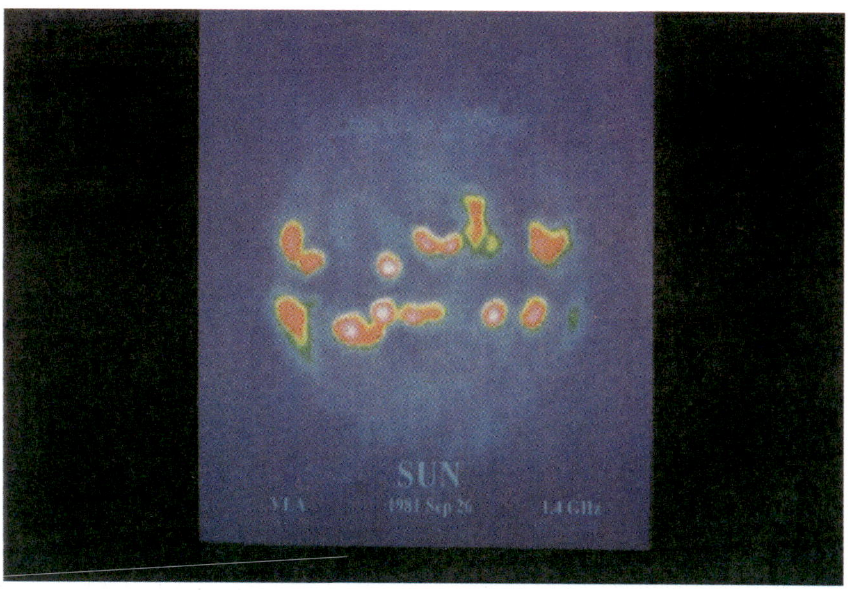

Plate 22 Brightness distribution of the sun. Observed by GA Dulk and DE Gary on September 26,1981, in a wavelength of 20 cm, using the VLA with all the 27 antennas. (*Courtesy*: NRAO)

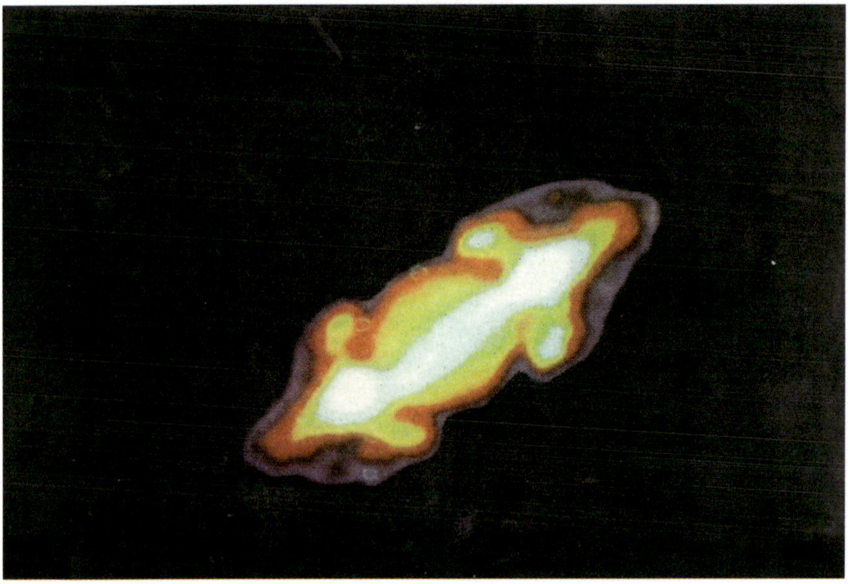

Plate 23 A radio picture of Jupiter. Observed by I de Pater in May 1981, in wavelength of 21 cm, using the VLA with all the 27 antennas. (*Courtesy*: NRAO)

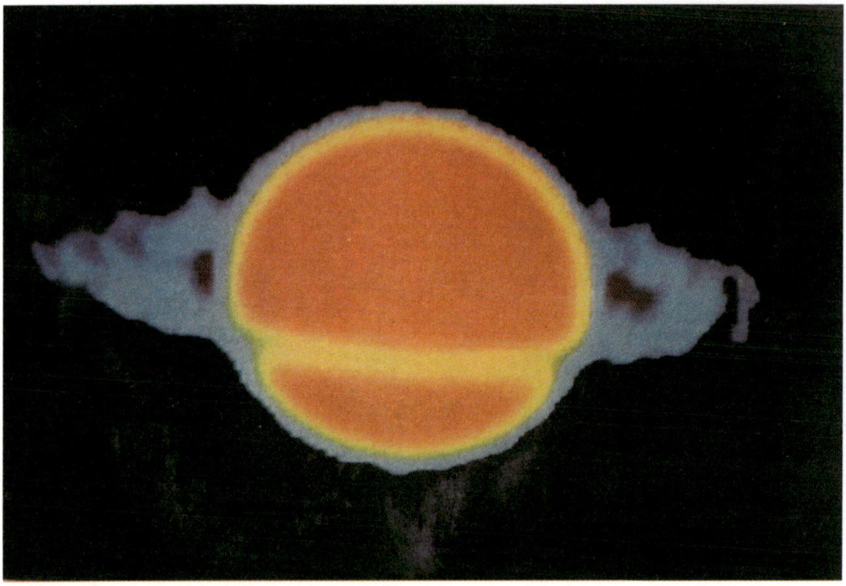

Plate 24 A radio picture of Saturn and its rings. Observed by I de Pater and JR Dickel on January 24-25, 1982, in a wavelength of 2 cm, using the VLA. (*Courtesy*: NRAO)

Plate 25 A dish of the antenna of the Giant Metrewave Radio Telescope, placed on ground. The GMRT is under construction. (*Courtesy*: NCRA/TIFR)